Aerospace Vehicle Dynamics Simulation:
Principles, Techniques, and Applications

BY:

Immanuel B. Gitamo, PhD

Contents

INTRODUCTION

The design, analysis, and operation of aerospace vehicles are inherently complex endeavors, demanding a rigorous understanding of multidisciplinary phenomena ranging from fluid dynamics and thermodynamics to control theory and structural mechanics. Central to this understanding is the ability to accurately predict a vehicle's dynamic behavior under a wide range of operating conditions and environmental influences. Physical prototypes and extensive flight testing, while essential for final validation, are often prohibitively expensive, time-consuming, and, in some cases, impossible due to the unique environments in which aerospace vehicles operate.

Consequently, modeling and simulation have become indispensable tools for aerospace engineers, providing a virtual laboratory where designs can be iterated, control strategies tested, and mission performance optimized before any physical hardware is committed.

This book, *Aerospace Vehicle Dynamics Simulation: Principles, Techniques, and Applications*, presents comprehensive treatment of the theory and practice of simulating the dynamic behavior of aerospace vehicles. Building from first principles, it provides a rigorous yet accessible introduction to the mathematical foundations underlying vehicle dynamics, emphasizing a clear and consistent framework for modeling various types of aerospace systems, including fixed-wing aircraft, rotary-wing aircraft, rockets, satellites, and unmanned aerial vehicles (UAVs).

The book adopts a systems engineering approach, guiding the reader through the entire simulation process, from model development and numerical integration to validation and uncertainty quantification. Key features of this comprehensive treatment include:

- **A Thorough Exposition of Core Principles:** Detailed coverage of coordinate systems and transformations, equations of motion for rigid bodies, aerodynamic and propulsion system modeling, and control system design.
- **Emphasis on Numerical Methods:** Detailed explanation of numerical integration techniques, including Euler, Runge-Kutta, and variable step-size methods, with practical guidance on selecting the appropriate method for a given application.
- **Practical Implementation Examples:** Numerous examples illustrating the application of theoretical concepts to real-world aerospace problems, with code examples in both MATLAB and Python.
- **Advanced Topics in Control and Simulation:** In-depth treatment of state-space control, robust control, adaptive control, hardware-in-the-loop testing, and real-time simulation techniques.
- **Uncertainty Quantification and Sensitivity Analysis:** Rigorous treatment of uncertainty quantification and sensitivity analysis techniques to assess the reliability and robustness of simulation results.
- **Emerging Trends and Future Directions:** Exploration of cutting-edge research areas, including machine learning, digital twins, advanced visualization, and the potential of quantum computing in aerospace simulation.

This book is intended for advanced undergraduate and graduate students in aerospace engineering, as well as practicing engineers seeking to enhance their knowledge and skills in aerospace vehicle dynamics simulation. It provides a solid foundation in the fundamental principles, practical techniques, and emerging trends that are shaping the future of aerospace engineering. By combining rigorous theory with practical applications, this book aims to empower readers to develop and deploy accurate, reliable, and insightful simulations that drive innovation in the design, analysis, and operation of next-generation aerospace vehicles.

Introduction to Aerospace Vehicle Dynamics: Why Simulate?

include the effects of angle of attack, wing flaps, and other factors.

Calculation Example:

Let's say we have an aircraft with the following characteristics:

- S = 20 m² (Wing area)
- C_L = 0.8 (Lift coefficient)
- V = 50 m/s (Airspeed)
- ρ = 1.225 kg/m³ (Air density at sea level)

Using the lift equation, we can calculate the lift force:

L = 0.5 * 1.225 kg/m³ * (50 m/s)² * 20 m² * 0.8 L = 24500 N (Newton's)

This simple calculation shows how we can use a model to estimate the lift force acting on the aircraft.

1.4 Simulation: Bringing the Model to Life

Simulation is the process of using a model to predict the behavior of a system over time. This typically involves using computer software to solve the equations that describe the model. The output of a simulation can be visualized in various ways, such as graphs, animations, or numerical data.

1.5 Types of Simulations

- **Static vs. Dynamic Simulations:** Static simulations represent the system at a single point in time, while dynamic simulations show how the system evolves over time. Aerospace vehicle dynamics generally involves dynamic simulations.
- **Linear vs. Nonlinear Simulations:** Linear simulations use linear equations to represent the system, while nonlinear

simulations use nonlinear equations. Most aerospace vehicle dynamics problems are inherently nonlinear.

- **Deterministic vs. Stochastic Simulations:** Deterministic simulations produce the same output for the same input, while stochastic simulations include random elements to account for uncertainty.
- **Hardware-in-the-Loop (HIL) Simulation:** This integrates physical hardware components (e.g., flight computers, actuators) into the simulation loop, allowing for more realistic testing.

1.6 The Modeling and Simulation Process: A General Overview

The modeling and simulation process typically involves the following steps:

1. **Define the Objectives:** What questions are we trying to answer with the simulation?
2. **Develop the Model:** Create a mathematical representation of the system, making appropriate simplifications and assumptions.
3. **Implement the Model:** Translate the mathematical model into computer code.
4. **Validate the Model:** Compare the simulation results with experimental data or other reliable sources to ensure that the model is accurate.
5. **Run the Simulation:** Execute the simulation and collect the results.
6. **Analyze the Results:** Interpret the simulation results and draw conclusions.
7. **Iterate and Refine:** Modify the model or simulation based on the results and repeat the process.

1.7 Software Tools for Aerospace Simulation

A wide range of software tools are available for aerospace simulation, including:

- **MATLAB/Simulink:** A widely used platform for modeling and simulating dynamic systems.
- **Python** (with libraries like NumPy, SciPy, and Matplotlib): A versatile programming language with powerful scientific computing capabilities.
- **X-Plane and FlightGear:** Flight simulator software that can be used for research and development.
- **Specialized CFD (Computational Fluid Dynamics) software:** Used for simulating airflow around aerospace vehicles.
- **STK (Systems Tool Kit):** Used for modeling and simulating space missions.

1.8 Looking Ahead: What We'll Cover

In the following chapters, we will delve deeper into the principles of modeling and simulating aerospace vehicle dynamics. We will cover topics such as:

- Coordinate systems and transformations
- Equations of motion for rigid bodies
- Aerodynamic modeling
- Propulsion system modeling
- Control system design
- Simulation techniques
- Case studies of aerospace vehicle simulations

1.9 Exercises

- Explain why simulation is important in aerospace vehicle design.
- Describe the difference between a model and a simulation.

- Give an example of a simplification that might be made when developing a model of an aircraft.
- Research and list three software tools commonly used for aerospace simulation.

Key Takeaways from this Chapter:

- Simulation is critical for designing, testing, and analyzing aerospace vehicles.
- A model is a simplified representation of a real-world system.
- Simulation involves using a model to predict the behavior of a system over time.
- The modeling and simulation process involves several steps, including defining objectives, developing the model, validating the model, and analyzing the results.
- A variety of software tools are available for aerospace simulation.

2.2 Defining Our Turf: Common Coordinate Systems in Aerospace

Let's explore the most used coordinate systems in aerospace engineering:

2.2.1 Inertial Frame (I-Frame): The Unmoving Observer

The Inertial Frame (I-Frame) is a theoretical, non-accelerating, and non-rotating reference frame. In essence, it's considered "fixed" in space. Why is this important? Newton's Laws of Motion are only valid in inertial frames. While a truly inertial frame is an idealization (as the universe itself is expanding and everything is in motion), we can approximate it for practical purposes.

Think of it as an observer floating far out in space, unaffected by any gravitational forces or accelerations. This frame serves as the ultimate reference for describing the absolute motion of our aerospace vehicle. It's where we define the true velocity and acceleration.

While theoretical, the practical usefulness of the inertial frame requires that we can accurately relate measurements in a real (non-inertial) frame to this ideal frame.

2.2.2 Earth-Centered Inertial (ECI) Frame: A Practical Approximation

For most Earth-based aerospace applications, we use the Earth-Centered Inertial (ECI) Frame. This frame is centered at the Earth's center of mass, and its axes point towards fixed stars (distant celestial objects). The ECI frame is not truly inertial (due to the Earth's orbit around the Sun and other celestial motions), but it's a good approximation for many simulations, especially those that don't span extremely long durations.

- The Z-axis points along the Earth's rotational axis, towards the North Pole.
- The X-axis points towards the vernal equinox (the point where the Sun crosses the celestial equator from south to north).
- The Y-axis completes the right-handed coordinate system.

The ECI frame is used extensively in satellite orbit calculations and interplanetary trajectory planning.

2.2.3 Earth-Centered, Earth-Fixed (ECEF) Frame: Ground Truth

The Earth-Centered, Earth-Fixed (ECEF) Frame is also centered at the Earth's center of mass, but unlike the ECI frame, it rotates with the Earth. This frame is extremely useful for defining locations on the Earth's surface. Coordinates in the ECEF frame are often expressed as X, Y, and Z Cartesian coordinates.

- The Z-axis points along the Earth's rotational axis, towards the North Pole (same as ECI).
- The X-axis points towards the Greenwich meridian (0 degrees longitude).
- The Y-axis completes the right-handed coordinate system.

GPS systems and mapping applications rely heavily on the ECEF frame to provide accurate location information. Latitude, longitude, and altitude are often derived from ECEF coordinates.

2.2.4 Body Frame (B-Frame): The Vehicle's Perspective

The Body Frame (B-Frame) is a crucial frame that is fixed to the aerospace vehicle itself. Its origin is typically located at the vehicle's center of gravity. The axes of the B-Frame are aligned with the vehicle's principal axes of inertia (axes about which the vehicle's mass is distributed most evenly).

- The X-axis typically points out the nose of the vehicle.
- The Y-axis typically points out the right wing.
- The Z-axis typically points downwards (completing the right-handed system).

The B-Frame is essential for describing the forces and moments acting on the vehicle, as well as the control inputs applied by the pilot or autopilot. For example, aerodynamic forces and control surface deflections are naturally expressed in the B-Frame.

2.2.5 Wind Frame (W-Frame): Riding the Air

The Wind Frame (W-Frame) is another essential coordinate system that is aligned with the relative wind (the direction of airflow relative to the vehicle). The x-axis of the W-Frame is aligned with the relative wind velocity vector.

- The X-axis points in the direction of the relative wind.
- The Z-axis is in the plane of symmetry of the vehicle and perpendicular to the X-axis.
- The Y-axis completes the right-handed coordinate system.

The W-Frame is particularly useful for analyzing aerodynamic forces and moments, as the lift and drag forces are naturally aligned with the axes of this frame. The angle of attack (the angle between the vehicle's longitudinal axis and the relative wind) and the sideslip angle (the angle between the vehicle's velocity vector and the plane of symmetry) are also defined with respect to the W-Frame.

2.2.6 Navigation Frame (N-Frame): The Pilot's Reference

The Navigation Frame (N-Frame) is a local coordinate system aligned with the local horizontal plane. This frame is often used in navigation systems to represent the aircraft's position and orientation relative to the Earth's surface. Depending on the application, the N-Frame can have different axis orientations. A typical choice would be:

- Z-axis pointed down toward the local gravity vector.
- X-axis pointed north.
- Y-axis pointed east.

2.3 The Art of Transformation: Bridging the Coordinate Gaps

The ability to seamlessly switch between these coordinate systems is crucial for a comprehensive understanding of aerospace vehicle dynamics. This is achieved through coordinate transformations, which are mathematical operations that convert coordinates or vectors from one frame to another. These transformations are often represented using rotation matrices.

2.3.1 Rotation Matrices: The Key to Orientation

Rotation matrices are 3×3 matrices that describe the orientation of one coordinate system relative to another. They represent rotation about a specific axis. The elements of a rotation matrix are trigonometric functions (sines and cosines) of the rotation angle.

A rotation matrix transforms a vector from one coordinate system to another by multiplying the vector by the rotation

quaternions to rotation matrices and vice versa will be discussed in a later chapter.

2.4 Putting It into Practice: Transformation Calculations

Let's solidify our understanding with a few practical examples:

- **Example 1: Transforming a Velocity Vector from Body Frame to Inertial Frame**

Suppose an aircraft has a velocity vector of v_B = [100, 5, −2] m/s in its body frame. The aircraft's Euler angles are ϕ = 5°, θ = 2°, and ψ = 1°. We want to find the velocity vector v_I in the inertial frame.

First, we calculate the rotation matrix R_{IB} using the Euler angles and the Roll-Pitch-Yaw sequence (as shown in the Python example earlier). Then, we multiply the velocity vector v_B by the rotation matrix R_{IB}:

v_I = R_{IB} v_B

The resulting vector v_I will represent the aircraft's velocity in the inertial frame.

Python Code:

Python Run

```
Import numpy as np
# Euler angles in degrees
phi = np.radians(5)
theta = np.radians(2)
psi = np.radians(1)
# Rotation matrices
Rx = np.array([[1, 0, 0], [0, np.cos(phi), -np.sin(phi)], [0, np.sin(phi), np.cos(phi)]])
Ry = np.array([[np.cos(theta), 0, np.sin(theta)], [0, 1, 0], [-np.sin(theta), 0, np.cos(theta)]])
Rz = np.array([[np.cos(psi), -np.sin(psi), 0], [np.sin(psi), np.cos(psi), 0], [0, 0, 1]])
# Rotation matrix from Body to Inertial
RIB = Rz @ Ry @ Rx
# Velocity vector in Body Frame
vB = np.array([100, 5, -2])
# Velocity vector in Inertial Frame
vI = RIB @ vB
print ("Velocity in Inertial Frame:", vI)
```

• Example 2: Transforming Coordinates from ECEF to Latitude, Longitude, and Altitude

Given a point with ECEF coordinates r_{ECEF} = [4000000, 1000000, 4900000] meters, we want to find its corresponding latitude, longitude, and altitude. This transformation is more complex and involves the Earth's ellipsoid model. While the full derivation is beyond the scope of this chapter, many libraries and online calculators can perform this conversion.

Key Steps:

1. Use the ECEF coordinates to calculate the geocentric latitude (latitude with respect to the Earth's center).
2. Use the geocentric latitude and ECEF coordinates to calculate the geodetic latitude (latitude with respect to the Earth's ellipsoid).
3. Calculate the longitude from the ECEF coordinates.
4. Calculate the altitude above the ellipsoid.

This calculation demonstrates how different coordinate systems are used for specific purposes. GPS devices perform these transformations constantly to provide us with our location information.

2.5 Summary and Look Ahead

This chapter has laid the groundwork for understanding the motion of aerospace vehicles by introducing the essential concepts of coordinate systems and transformations. We have explored several common coordinate systems used in aerospace engineering, including inertial, ECI, ECEF, body, and wind frames. We have also discussed rotation matrices, Euler angles, and the importance of coordinate transformations.

Understanding these concepts is crucial for the upcoming chapters, where we will delve into the equations of motion for rigid bodies and learn how to simulate the dynamics of aerospace vehicles.

2.6 Exercises

1. Explain the purpose of a coordinate system in aerospace vehicle dynamics.
2. Describe the differences between the inertial frame and the Earth-Centered Inertial (ECI) frame.
3. What are Euler angles? Explain the Roll-Pitch-Yaw sequence and why the order matters.
4. Why are quaternions sometimes preferred over Euler angles for representing rotations?
5. Write a Python function that takes Euler angles (in degrees) as input and returns the corresponding rotation matrix (using the Roll-Pitch-Yaw sequence).
6. An aircraft is flying with a heading of 45 degrees (yaw), a pitch angle of 10 degrees, and a roll angle of 2 degrees. A sensor on the aircraft measures an acceleration of [1, 0, 0] m/s² in the body frame. What is the acceleration vector in the inertial frame? Provide your code (Python preferred) and the result.
7. Research and describe one real-world scenario where accurately transforming between different coordinate systems is essential for a successful aerospace mission.

- **Newton's Second Law:** The sum of the forces acting on an object is equal to the mass of that object multiplied by its acceleration (F = ma).
- **Newton's Third Law:** For every action, there is an equal and opposite reaction.

While seemingly simple, these laws provide the fundamental relationships between force, mass, and acceleration that govern the motion of all objects, including aerospace vehicles.

3.3 Forces and Moments: The Actors of Motion

Forces and moments are the agents that cause changes in the linear and angular momentum of an aerospace vehicle. Let's examine the main players:

- **Forces:** Forces are external influences that tend to cause a body to accelerate linearly. Key forces acting on an aerospace vehicle include:
- **Gravity (Weight):** The force of attraction between the vehicle and the Earth (or other celestial body).
- **Aerodynamic Forces:** Forces generated by the flow of air around the vehicle, including lift, drag, and side force.
- **Thrust:** The force generated by the vehicle's engines or rockets.
- **Control Surface Forces:** Forces generated by deflecting control surfaces (e.g., ailerons, elevators, rudders).

- **Moments (Torques):** Moments are external influences that tend to cause a body to accelerate angularly. They are the rotational equivalent of forces. Key moments acting on an aerospace vehicle include:

- **Aerodynamic Moments:** Moments generated by the flow of air around the vehicle, including pitching moment, rolling moment, and yawing moment.
- **Control Surface Moments:** Moments generated by deflecting control surfaces.
- **Engine Moments:** Moments generated by asymmetric thrust or engine placement.

3.4 Linear Equations of Motion: Translating Through Space

Let **F** be the vector sum of all external forces acting on the aerospace vehicle, and let **a_c** be the acceleration of the vehicle's center of gravity (CG) with respect to an inertial frame. According to Newton's Second Law, we have:

F = m a_c

Where:

- **F** = [F_x, F_y, F_z] ^T is the vector sum of all external forces acting on the vehicle.
- **m** is the mass of the vehicle.
- **a_c** = [a_{cx}, a_{cy}, a_{cz}]^T is the acceleration vector of the vehicle's center of gravity with respect to an inertial frame.

This simple equation represents the fundamental relationship between force and linear acceleration. However, for aerospace vehicles, we often need to express these quantities in different coordinate systems.

If we express the forces and acceleration in the body frame (B-Frame), then the equation becomes:

F_B = m a_{cB}

Where **F_B** and **a_{cB}** are the force and acceleration vectors, respectively, expressed in the body frame.

To relate the acceleration in the body frame to the acceleration in an inertial frame, we need to consider the effects of the vehicle's rotation. Let **v_I** be the velocity of the vehicle in the inertial frame, and let **ω_B** = [p, q, r]^T be the angular velocity vector of the body frame with respect to the inertial frame, expressed in the body frame (p, q, and r are the roll, pitch, and yaw rates, respectively). Then, the acceleration of the vehicle in the inertial frame is given by:

a_I = d/dt (v_I) = d/dt (R_{IB} v_B)

Where **R_{IB}** is the rotation matrix that transforms vectors from the body frame to the inertial frame. Expanding the derivative using the product rule, we get:

a_I = (dR_{IB}/dt) v_B + R_{IB} (dv_B/dt)

It can be shown that:

dR_{IB}/dt v_B = R_{IB} (ω_B × v_B)

Therefore,

a_I = R_{IB} (dv_B/dt + ω_B × v_B)

Let **a_{cB} = dv_B/dt** be the linear acceleration of the vehicle in the body frame. Then the above equation can be rewritten as:

R_{IB} a_{cB} = a_I – R_{IB} (ω_B × v_B)

Therefore,

a_{cB} = R_{BI} a_I – (ω_B × v_B)

And

F_B / m = R_{BI} a_I – (ω_B × v_B)

This last equation relates the forces in the body frame to the acceleration of the vehicle with respect to an inertial frame. This accounts for the rotation of the vehicle.

3.5 Angular Equations of Motion: Rotating in Space

The angular equations of motion describe the relationship between the moments acting on a vehicle and its angular acceleration. The rotational equivalent of Newton's Second Law is:

M = I α

Where:

- **M = [L, M, N]^T** is the vector sum of all external moments acting on the vehicle (L = rolling moment, M = pitching moment, N = yawing moment).
- **I** is the inertia tensor (a 3×3 matrix representing the vehicle's resistance to angular acceleration).
- **α = [α_x, α_y, α_z]^T** is the angular acceleration vector.

However, this simple equation is not sufficient for describing the complex rotational motion of aerospace vehicles.

• I_{xx}, I_{yy}, I_z_x, and I_xz are the moments and products of inertia.

These equations form a coupled set of nonlinear differential equations. Solving them requires numerical integration techniques, which we will discuss in a later chapter.

3.8 Simplifications and Assumptions: Balancing Accuracy and Complexity

The equations of motion presented above are relatively general. However, in many cases, we can make simplifying assumptions to reduce the complexity of the equations. Common assumptions include:

• **Rigid Body Assumption:** The vehicle is assumed to be a perfectly rigid body, meaning that it does not deform under the influence of external forces.

• **Symmetry Assumption:** The vehicle is assumed to be symmetrical about the xz-plane, which simplifies the inertia tensor ($I_{xy} = I_yz = 0$).

• **Small Angle Assumption:** For small angles of attack and sideslip, we can approximate trigonometric functions using linear approximations (e.g., $\sin(\theta) \approx \theta$, $\cos(\theta) \approx 1$).

These assumptions can significantly simplify the equations of motion, making them easier to solve and analyze. However, it is important to be aware of the limitations of these assumptions and to ensure that they are valid for the specific application.

3.9 Calculation Examples

• **Example 1: Calculating Forces Required for a Constant Turn**

Assume an airplane is performing a coordinated, level turn at a constant speed of 200 m/s. The bank angle (roll angle) is 30 degrees. The mass of the airplane is 5000 kg. Calculate the lift required to maintain the turn.

The required lift is **$L = mg / \cos(\varphi)$**, where φ is the bank angle. Therefore,
$L = 5000 \times 9.81 / \cos(30°) = 56{,}697$ N

- **Example 2: Effect of Gyroscopic Moments**

An aircraft has **$I_{xx} = 1000$ kg·m², $I_{yy} = 2000$ kg·m², $Izz = 2500$ kg·m²**. It is pitching at a rate of **$q = 0.1$ rad/s** and yawing at a rate of **$r = 0.05$ rad/s**. Calculate the rolling moment generated by the gyroscopic effect.

$L = (Izz - I_{yy})\, q\, r = (2500 - 2000) \times 0.1 \times 0.05 = 2.5$ N·m

3.10 Numerical Integration

Solving these equations almost always requires numerical integration. We will cover this in greater detail in the following chapters. Methods such as the Euler Method, Runge–Kutta (RK4), and others are frequently used.

3.11 Summary and Conclusion

This chapter has provided a comprehensive overview of the equations of motion for rigid aerospace vehicles. We have discussed Newton's Laws of Motion, forces and moments, linear and angular equations of motion, the inertia tensor, and simplifying assumptions. Understanding these concepts is crucial for simulating the behavior of aerospace vehicles and designing effective control systems. In the following chapters, we will build upon this foundation to explore advanced topics such as aerodynamic modeling, control system design, and simulation techniques.

3.12 Exercises

1. Explain the difference between a force and a moment. Give examples of each that are relevant to aerospace vehicles.
2. Derive the linear equations of motion for a simplified case where the vehicle is only allowed to move in one dimension (e.g., vertical motion).
3. What is the inertia tensor? Explain its significance in the angular equations of motion.
4. Research and describe a real-world scenario where gyroscopic moments have a significant impact on the behavior of an aerospace vehicle.
5. An aircraft has an inertia tensor with $\mathbf{I_{xx} = 5000\ kg{\cdot}m^2}$, $\mathbf{I_{yy} = 8000\ kg{\cdot}m^2}$, and $\mathbf{Izz = 10{,}000\ kg{\cdot}m^2}$. It is rolling at a rate of $\mathbf{p = 0.2\ rad/s}$ and pitching at a rate of $\mathbf{q = 0.1\ rad/s}$. Assuming that the products of inertia are negligible, calculate the yawing moment generated by the gyroscopic effect.

Aerodynamic Modeling: Taming the Winds

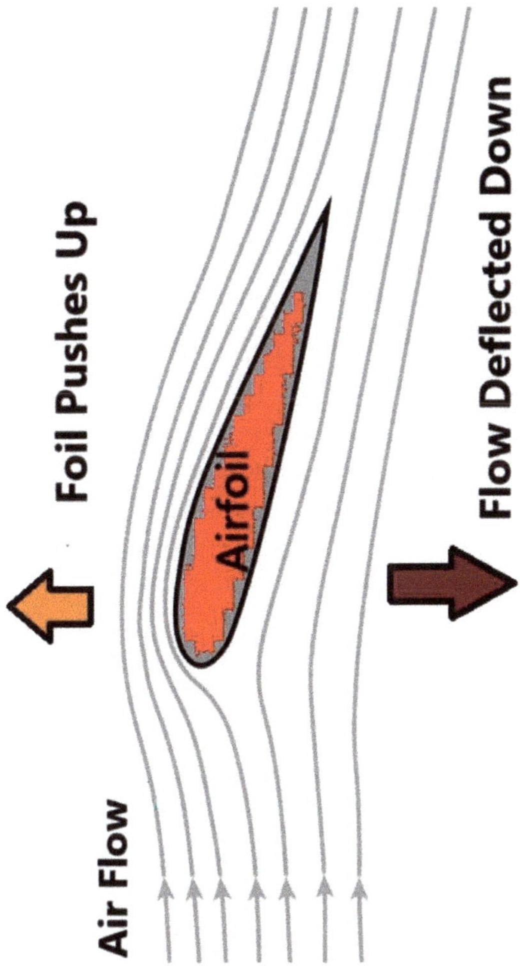

This chapter provided a comprehensive exploration of aerodynamic modeling, providing readers with the necessary knowledge and skills to develop accurate and reliable aerodynamic models for aerospace vehicle simulations.

4.1 Introduction: The Dance between Air and Vehicle

Aerodynamics, the study of how air moves around objects, is a cornerstone of aerospace vehicle dynamics. The aerodynamic forces and moments generated by this airflow are the primary drivers of an aircraft's flight, impacting its lift, drag, stability, and control. Accurately modeling these aerodynamic effects is therefore essential for designing, analyzing, and simulating aerospace vehicles.

In this chapter, we will delve into the world of aerodynamic modeling, exploring different approaches for representing these complex phenomena. We will cover topics such as aerodynamic coefficients, airfoil theory, wing theory, and the influence of various factors on aerodynamic performance.

4.2 Aerodynamic Forces and Moments: The Key Players

As discussed in Chapter 3, forces and moments dictate the motion of aerospace vehicles. Aerodynamic forces and moments arise from the pressure and shear stress distribution acting on the vehicle's surface due to the airflow. The key aerodynamic forces are:

- **Lift (L):** The force acting perpendicular to the relative wind, primarily responsible for supporting the vehicle's weight and enabling flight.

- **Drag (D):** The force acting parallel to the relative wind, opposing the vehicle's motion through the air.
- **Side Force (Y):** The force acting perpendicular to both the lift and drag forces, typically caused by sideslip or asymmetric airflow.

The key aerodynamic moments are:

- **Rolling Moment (L):** The moment acting about the vehicle's longitudinal axis, causing it to roll.
- **Pitching Moment (M):** The moment acting about the vehicle's lateral axis, causing it to pitch up or down.
- **Yawing Moment (N):** The moment acting about the vehicle's vertical axis, causing it to yaw (turn).

4.3 Aerodynamic Coefficients: Quantifying the Invisible

Aerodynamic coefficients are dimensionless numbers that relate the aerodynamic forces and moments to the vehicle's geometry, airspeed, and air density. They provide a convenient way to characterize the aerodynamic performance of a vehicle and allow for comparisons between different designs.

The most commonly used aerodynamic coefficients are:

- **Lift Coefficient (C_L):**
  C_L = L / (0.5 · ρ · V^2 · S)
- **Drag Coefficient (C_D):**
  C_D = D / (0.5 · ρ · V^2 · S)
- **Side Force Coefficient (C_Y):**
  C_Y = Y / (0.5 · ρ · V^2 · S)
- **Rolling Moment Coefficient (C_l):**
  C_l = L / (0.5 · ρ · V^2 · S · b)
- **Pitching Moment Coefficient (C_m):**
  C_m = M / (0.5 · ρ · V^2 · S · c)

- **Yawing Moment Coefficient (C_n):** C_n = N / (0.5 · ρ · V² · S · b)

  Where:

L, D, and Y are the lift, drag, and side forces, respectively.

• L, M, and N are the rolling, pitching, and yawing moments, respectively.

- ρ is the air density.
- V is the airspeed.
- S is the wing area.
- b is the wingspan.
- c is the mean aerodynamic chord (MAC).

These coefficients are typically functions of the angle of attack (α), sideslip angle (β), control surface deflections (δ_a, δ_e, δ_r), and Mach number (Ma).

4.4 Airfoil Theory: Understanding Lift Generation

An airfoil is the cross-sectional shape of a wing. Airfoil theory provides a fundamental understanding of how lift is generated.

4.4.1 Pressure Distribution:

Lift is generated due to a pressure difference between the upper and lower surfaces of the airfoil. The airflow over the upper surface travels a longer distance than the airflow over the lower surface, resulting in a lower pressure on the upper surface and a higher pressure on the lower surface. This pressure difference creates an upward force, which is lift.

4.4.2 Angle of Attack:

The angle of attack (α) is the angle between the airfoil's chord line (an imaginary line connecting the leading and trailing edges) and the relative wind. As the angle of attack increases, the pressure difference between the upper and lower surfaces increases, resulting in increased lift. However, beyond a certain angle of attack (the stall angle), the airflow separates from the upper surface, leading to a sudden loss of lift.

4.4.3 Airfoil Characteristics:

Different airfoil shapes have different lift and drag characteristics. Some airfoils are designed for high lift, while others are designed for low drag. The choice of airfoil depends on the specific application.

- **Symmetric Airfoils:** Have identical upper and lower surfaces, generating zero lift at zero angle of attack. Often used for control surfaces.
- **Cambered Airfoils:** Have a curved upper surface, generating lift even at zero angle of attack. Common for wing designs.

4.4.4 Thin Airfoil Theory:

This simplified theory provides a relationship between the lift coefficient and the angle of attack for thin airfoils:

C_L ≈ 2π α

This equation is valid for small angles of attack and thin airfoils.

4.4.5 Computational Fluid Dynamics (CFD):

CFD simulations provide detailed information about the airflow around airfoils and wings, including pressure distributions, velocity fields, and turbulence.

4.5 Wing Theory: From Airfoil to 3D Wing

While airfoil theory provides insights into lift generation, it only considers a two-dimensional cross-section. Wing theory extends these concepts to three-dimensional wings.

4.5.1 Finite Wing Effects:

Unlike airfoils, wings have a finite span. This leads to the formation of wingtip vortices, which are swirling masses of air that trail behind the wingtips. These vortices induce a downward velocity component (downwash) that reduces the effective angle of attack of the wing.

4.5.2 Induced Drag:

The downwash induced by the wingtip vortices creates an additional drag force called induced drag. Induced drag is proportional to the square of the lift coefficient and is inversely proportional to the wing's aspect ratio (the ratio of wingspan to chord).

4.5.3 Aspect Ratio:

A high aspect ratio wing (long and slender) has lower induced drag than a low aspect ratio wing (short and stubby). Gliders typically have high aspect ratio wings to maximize lift-to-drag ratio.

4.5.4 Wing Planform:

The shape of the wing (planform) also affects its aerodynamic characteristics. Common wing planforms include rectangular, elliptical, tapered, and swept wings.

4.5.5 Lifting-Line Theory:

A simplified theory that accounts for the finite span effects of wings. It represents the wing as a line of vortices, allowing for the calculation of the induced drag and lift distribution.

4.5.6 Winglets:

Small vertical surfaces attached to the wingtips that reduce induced drag by disrupting the formation of wingtip vortices.

4.6 Modeling Techniques: Bridging Theory and Simulation

Several modeling techniques are used to represent aerodynamic forces and moments acting on aerospace vehicles. The choice of technique depends on the desired accuracy and computational cost.

- **Control Surface Deflections (δ_a, δ_e, δ_r):** Control surfaces generate aerodynamic forces and moments that control the vehicle's attitude and flight path.
- **High-Lift Devices:** Flaps and slats increase wing lift at low speeds, improving takeoff and landing performance.

4.8 Modeling Control Surfaces

Control surfaces (ailerons, elevators, rudder) are essential for controlling an aerospace vehicle's attitude and flight path. Accurate modeling of their effects is crucial for control system design.

The influence of control surfaces on aerodynamic coefficients can be represented using **stability derivatives** or **look-up tables**. For example, the lift coefficient can be expressed as:

C_L = C_{L0} + C_{Lα} α + C_{Lδe} δ_e

Where:

- **C_{L0}** is the lift coefficient at zero angle of attack and zero elevator deflection.
- **C_{Lα}** is the lift curve slope (change in lift coefficient per unit angle of attack).
- **C_{Lδe}** is the elevator effectiveness (change in lift coefficient per unit elevator deflection).
- **δ_e** is the elevator deflection angle.

4.9 Calculation Examples

Example 1: Calculating Lift and Drag

An aircraft is flying at 100 m/s at sea level (ρ = 1.225 kg/m³). The wing area is 20 m², and the lift and drag coefficients are C_L = 0.5 and C_D = 0.05.

L=0.5·ρ·V2·S·CL=0.5·1.225·1002·20·0.5=61,250 NL = 0.5 \cdot \rho \cdot V^2 \cdot S \cdot C_L = 0.5 \cdot 1.225 \cdot 100^2 \cdot 20 \cdot 0.5 = 61,250 \, \text{N}L=0.5·ρ·V2·S·CL =0.5·1.225·1002·20·0.5=61,250N D=0.5·ρ·V2·S·CD=0.5·1.225·1002·20·0.05=6,125 ND = 0.5 \cdot \rho \cdot V^2 \cdot S \cdot C_D = 0.5 \cdot 1.225 \cdot 100^2 \cdot 20 \cdot 0.05 = 6,125 \, \text{N}D=0.5·ρ·V2·S·CD =0.5·1.225·1002·20·0.05=6,125N

Example 2: Calculating Induced Drag

A wing has a wingspan of 10 m and a wing area of 20 m². Calculate the aspect ratio and the induced drag coefficient for C_L = 0.8. Assume Oswald efficiency factor e = 0.85.

Aspect Ratio (AR) = b² / S = (10 m)² / 20 m² = 5 C_{Di} = C_L² / (π * AR * e) = (0.8)² / (π * 5 * 0.85) = 0.0477

4.10 Simulation and Validation

Once an aerodynamic model is developed, it must be **validated** against experimental data or CFD simulations. This ensures the model accurately represents the vehicle's aerodynamic characteristics. Validation typically involves comparing predicted forces, moments, and coefficients against measurements for various flight conditions.

4.11 Conclusion

Accurate aerodynamic modeling is essential for simulating aerospace vehicle behavior and designing effective control systems. This chapter covered:

- Aerodynamic coefficients and their role in quantifying lift, drag, and moments.
- Airfoil theory and the principles of lift generation.
- Wing theory, including finite wing effects, induced drag, and aspect ratio.
- Modeling techniques, from look-up tables to CFD.
- Factors affecting aerodynamic performance and the modeling of control surfaces.

Mastering these concepts allows engineers to develop reliable simulations, guiding the design and analysis of aerospace vehicles.

4.12 Exercises

1. Explain the difference between lift and drag.
2. What are aerodynamic coefficients, and why are they useful?
3. Describe the key factors that affect the aerodynamic performance of a wing.
4. What is induced drag, and how can it be reduced?
5. Research and compare the advantages and disadvantages of different aerodynamic modeling techniques (look-up tables, empirical models, VLM, CFD).
6. An aircraft is flying at 150 m/s at an altitude of 5000 m (ρ = 0.736 kg/m^3). The wing area is 25 m^2, and the lift coefficient is C_L = 0.6. Calculate the lift force acting on the aircraft.
7. Using an online airfoil database (e.g., Airfoil Tools), research different airfoil shapes. Compare and contrast the lift and drag performance of a symmetric airfoil versus a cambered airfoil.

Propulsion System Modeling: The Engines of Flight

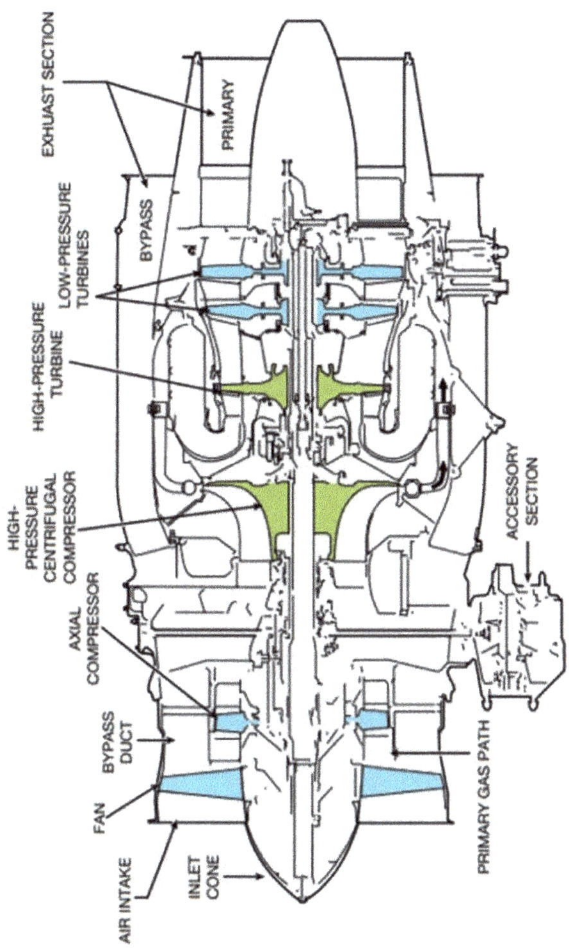

This chapter provides a comprehensive exploration of propulsion system modeling, equipping readers with the necessary knowledge and skills to develop accurate and reliable models for aerospace vehicle simulations.

5.1 Introduction: The Heart of Aerospace Vehicles

While aerodynamics shape the way air interacts with a vehicle, it's the propulsion system that provides the thrust necessary to overcome drag, generate lift (in the case of powered lift), and propel the vehicle through the atmosphere or space. Understanding how these systems work and accurately modeling their behavior is crucial for simulating and analyzing the performance of aerospace vehicles.

A poorly modeled propulsion system can lead to inaccurate predictions of range, endurance, climb rate, and overall flight dynamics.

This chapter will explore the fundamentals of aerospace propulsion systems, covering different engine types, performance parameters, and modeling techniques. We'll delve into the intricacies of reciprocating engines, gas turbines (turbojets, turbofans, turboprops), rocket engines (chemical and electric), and emerging propulsion technologies.

5.2 Propulsion System Fundamentals: Converting Energy into Thrust

All propulsion systems, regardless of their specific design, share a common goal: to generate thrust by accelerating a mass of working fluid. This process is governed by Newton's Third Law of Motion (action and reaction). The engine expels a mass of fluid (typically air or exhaust gases) in one direction, and the resulting reaction force propels the vehicle in the opposite direction.

The fundamental equation for thrust (T) is:

$T = \text{m-dot} * (V_e - V_0) + A_e * (P_e - P_0)$

Where:

- M-dot is the mass flow rate of the working fluid.
- V_e is the exhaust velocity.
- V_0 is the inlet velocity (the vehicle's airspeed).
- A_e is the exhaust nozzle area.
- P_e is the exhaust pressure.
- P_0 is the ambient pressure.

The first term (m-dot * (V_e - V_0)) represents the momentum thrust, which is the primary component of thrust for most propulsion systems. The second term (A_e * (P_e - P_0)) represents the pressure thrust, which is significant when the exhaust pressure is different from the ambient pressure (particularly in rocket engines).

5.3 Key Propulsion System Parameters

Several parameters are used to characterize the performance of propulsion systems:

• Thrust (T): The propulsive force generated by the engine.
• Specific Thrust (F/m-dot): The thrust produced per unit mass flow rate.

• Thrust-to-Weight Ratio (T/W): The ratio of the engine's thrust to its weight. A higher T/W ratio is desirable for high-performance aircraft and rockets.

• Specific Fuel Consumption (SFC): The amount of fuel consumed per unit of thrust per unit of time. A lower SFC is desirable for maximizing range and endurance. It's typically measured in kg/N/s or lb/lbf/hr.

• Specific Impulse (I_{sp}): A measure of the engine's efficiency in terms of thrust produced per unit weight flow rate of propellant. It's typically measured in seconds (s). Higher I_{sp} means greater efficiency in converting propellant into thrust. Important primarily for rocket engines.

$I_{sp} = T / (\text{m-dot} * g_0)$ (where g_0 is the standard gravity ≈ 9.81 m/s^2)

• Propulsive Efficiency (η_p): The ratio of the useful power delivered to the vehicle to the rate of kinetic energy addition to the propellant. It is a measure of how effectively the engine converts the energy of the fuel into useful thrust.

5.4 Reciprocating Engines: The Classic Power plant

Reciprocating engines, also known as piston engines, are internal combustion engines that use pistons to convert the chemical energy of fuel into mechanical work. They are commonly used in general aviation aircraft and drones due to their relatively low cost and simplicity.

5.4.1 Operating Principle:

Reciprocating engines operate on a four-stroke cycle (intake, compression, combustion, exhaust).

1. **Intake:** The piston moves down, drawing a mixture of air and fuel into the cylinder.
2. **Compression:** The piston moves up, compressing the air–fuel mixture.
3. **Combustion:** The compressed mixture is ignited by a spark plug, creating a high-pressure explosion that pushes the piston down.
4. **Exhaust:** The piston moves up, pushing the exhaust gases out of the cylinder.

5.4.2 Modeling Reciprocating Engines:

Reciprocating engine models can range from simple to highly complex.

Simple Model:

A simple model may represent the engine as a source of torque that is proportional to the throttle setting and engine speed:

Torque (τ) = K_t * Throttle * RPM

Where:

- K_t is a torque constant.
- Throttle is a value from 0 to 1 representing the throttle position.
- RPM is the engine speed (revolutions per minute).

The power output (P) can then be calculated as:

$P = \tau * \omega$

Where ω is the angular velocity (in radians per second).

More Complex Model:

A more complex model may include the effects of air density, fuel–air ratio, and engine temperature. These models often use look-up tables or empirical equations to represent the engine's performance characteristics. A table can relate brake horsepower (BHP) to manifold pressure, RPM, and fuel flow.

5.4.3 Key Parameters:

• Brake Horsepower (BHP): The power delivered by the engine to the propeller shaft.

• Manifold Pressure: The pressure of the air–fuel mixture in the intake manifold.

• Fuel Flow: The rate at which fuel is consumed by the engine.

• RPM: The engine speed (revolutions per minute).

5.5 Gas Turbine Engines: The Workhorses of Aviation

Gas turbine engines are internal combustion engines that use a turbine to extract energy from hot, high-pressure gases. They are widely used in commercial and military aircraft due to their high power-to-weight ratio and fuel efficiency. There are several types of gas turbine engines, including:

5.5.1 Turbojet Engines:

• **Operating Principle:**

Turbojet engines consist of an inlet, a compressor, a combustion chamber, a turbine, and a nozzle. Air enters the inlet, is compressed by the compressor, mixed with fuel in the combustion chamber, and ignited. The hot gases expand through the turbine, which extracts energy to drive the compressor. The remaining energy is used to accelerate the exhaust gases through the nozzle, generating thrust.

• **Modeling Turbojet Engines:**

Turbojet engines are typically modeled using performance maps that relate thrust, fuel flow, and other parameters to engine speed, altitude, and Mach number. These maps can be obtained from engine manufacturers or from experimental data.

• **Key Parameters:**

• Turbine Inlet Temperature (TIT): The temperature of the gases entering the turbine.

• Compressor Pressure Ratio (CPR): The ratio of the pressure at the compressor outlet to the pressure at the compressor inlet.

• Net Thrust: The total thrust produced by the engine.

5.5.2 Turbofan Engines:

• **Operating Principle:**

Turbofan engines are similar to turbojet engines, but they include a large fan at the front of the engine that bypasses some of the air around the core engine. This bypass air provides additional thrust and improves fuel efficiency.

• **Modeling Turbofan Engines:**

Turbofan engines are modeled similarly to turbojet engines, but the model must also account for the effects of the bypass air.

• **Key Parameters:**

Bypass Ratio (BPR): The ratio of the mass flow rate of the bypass air to the mass flow rate of the core engine air. A higher BPR generally results in improved fuel efficiency but reduced thrust at high speeds.

5.5.3 Turboprop Engines:

• **Operating Principle:**

Turboprop engines use a turbine to drive a propeller, which provides most of the thrust. The exhaust gases also provide some thrust, but to a lesser extent than in turbojet or turbofan engines.

• **Modeling Turboprop Engines:**

Turboprop engines are modeled by considering both the propeller thrust and the exhaust thrust. The propeller thrust is typically calculated using blade element theory or propeller performance maps.

• **Key Parameters:**

• Propeller Efficiency: The ratio of the useful power delivered by the propeller to the power input to the propeller.

• Propeller RPM: The rotational speed of the propeller.

• Torque: The torque delivered to the propeller shaft.

5.6 Rocket Engines: Powering Spaceflight

Rocket engines generate thrust by expelling hot gases produced by the combustion of propellants. Unlike air-breathing engines, rocket engines carry their own oxidizer, allowing them to operate in the vacuum of space.

5.6.1 Chemical Rocket Engines:

• **Operating Principle:**

Chemical rocket engines use the chemical energy stored in propellants to generate thrust. The propellants are typically a fuel (e.g., liquid hydrogen, kerosene) and an oxidizer (e.g., liquid oxygen, nitric acid). The propellants are mixed and ignited in a combustion chamber, producing hot, high-pressure gases that are expelled through a nozzle.

• **Modeling Chemical Rocket Engines:**

Chemical rocket engines are typically modeled using thermodynamic equations and nozzle flow calculations. The model must account for the combustion process, the properties of the propellants, and the geometry of the nozzle.

• **Key Parameters:**

• Chamber Pressure (P_c): The pressure in the combustion chamber.

• Nozzle Expansion Ratio (ε): The ratio of the nozzle exit area to the nozzle throat area.

• Propellant Mass Flow Rate: The rate at which propellants are consumed by the engine.
• Oxidizer-to-Fuel Ratio (O/F): The ratio of the mass flow rate of the oxidizer to the mass flow rate of the fuel. This ratio impacts combustion efficiency.

• **Basic Chemical Rocket Engine Thrust Equation:**

T = m-dot * V_e + A_e * (P_e - P₀)

Where:

• V_e is often approximated as V_e ≈ √(2

* h_c), where h_c is the change in enthalpy from combustion. In reality, this is determined using gas-dynamics equations involving chamber pressure, temperature, and nozzle expansion ratio.

✓ **Specific Impulse Calculation:**

The I_{sp} of a rocket engine is a critical indicator of its performance.
I_{sp} = V_e / g₀

Where:

• g₀ is standard gravity

• Example: A rocket engine with an exhaust velocity of 3000 m/s has a specific impulse of approximately 306 seconds.

5.6.2 Electric Propulsion Systems:

• **Operating Principle:**

Electric propulsion systems use electrical energy to accelerate a propellant. They offer much higher specific impulse than chemical rocket engines but produce lower thrust levels.

• **Types of Electric Propulsion Systems:**

• Ion Thrusters: Ionize a propellant (e.g., xenon) and accelerate the ions using an electric field.

• Hall Effect Thrusters: Use a magnetic field to confine electrons and accelerate ions.

• Arcjets: Heat a propellant using an electric arc and expand the hot gas through a nozzle.

• **Modeling Electric Propulsion Systems:**

Electric propulsion systems are modeled by considering the electrical power input, the propellant mass flow rate, and the thrust produced. The model must also account for the efficiency of the ionization and acceleration processes.

• **Key Parameters:**

• Thrust Power (P_T): The power associated with the thrust.

• Beam Voltage: The voltage used to accelerate the ions.

• Ionization Efficiency: The fraction of propellant atoms that are ionized.

• Propellant Utilization Efficiency: The fraction of propellant that is successfully converted into thrust.

5.7 Modeling Techniques: From Simple to Sophisticated

Several modeling techniques can be used to represent propulsion systems in aerospace vehicle simulations.

5.7.1 Simple Look-Up Tables:

This approach involves storing the engine's performance characteristics (thrust, fuel flow) as a function of throttle setting, altitude, and Mach number in a table. During the simulation, the engine's performance is interpolated from the table based on the current flight conditions. This method is computationally efficient but may not accurately capture complex engine dynamics.

5.7.2 Empirical Models:

These models use empirical equations to represent the engine's performance characteristics. The coefficients in these equations are typically determined from experimental data or engine manufacturer specifications.

5.7.3 Physics-Based Models:

These models use thermodynamic equations and fluid dynamics principles to simulate the engine's internal processes. Physics-based models are more accurate than empirical models but are also more computationally expensive. They are often used for detailed engine design and analysis.

5.7.4 Simulink Models:

Simulink provides a graphical environment for modeling and simulating dynamic systems. Propulsion systems can be modeled in Simulink using blocks that represent different engine components (e.g., compressor, turbine, and combustion chamber). Simulink models can be easily integrated with other vehicle models.

5.8 Integration with Flight Dynamics Models

The propulsion system model must be integrated with the flight dynamics model to accurately simulate the vehicle's behavior. This involves calculating the thrust force and applying it to the vehicle's center of gravity, as described in Chapter 3. The thrust force can be resolved into components in the body frame or the inertial frame, depending on the chosen coordinate system.

The thrust also affects the vehicle's mass, as fuel is consumed. The mass of the vehicle must be updated at each time step based on the fuel flow rate.

5.9 Factors Affecting Propulsion System Performance

Several factors can affect the performance of propulsion systems:

• **Altitude:** Air density decreases with altitude, reducing the thrust produced by air-breathing engines.

• **Mach Number:** At high Mach numbers, compressibility effects become significant, affecting the engine's performance.

• **Ambient Temperature:** Ambient temperature affects the density and pressure of the air entering the engine.

• **Throttle Setting:** The throttle setting controls the amount of fuel supplied to the engine, which affects the thrust produced.

• **Engine Health:** Engine degradation (e.g., wear and tear) can reduce the engine's performance.

5.10 Emerging Propulsion Technologies

Several emerging propulsion technologies are being developed for future aerospace vehicles:

• **Hypersonic Propulsion:** Scramjet engines are being developed for hypersonic flight (Mach 5 and above).

• **Electric Aircraft:** Electric propulsion systems are being explored for smaller aircraft and urban air mobility applications.

• **Nuclear Propulsion:** Nuclear thermal and nuclear electric propulsion systems are being developed for long-duration space missions.

5.11 Calculation Examples

- **Example 1:**

Calculating Thrust of a Turbojet Engine

A turbojet engine has a mass flow rate of 50 kg/s, an exhaust velocity of 600 m/s, and an inlet velocity of 250 m/s. The exhaust pressure is equal to the ambient pressure. Calculate the thrust produced by the engine.

T = m-dot * (V_e - V₀) = 50 kg/s * (600 m/s - 250 m/s) = 17,500 N

- **Example 2:**

Control System: Piloting Virtual Aircraft

In this chapter, we delve into the core principles of control system design, providing readers with a strong foundation in the techniques used to steer and stabilize aerospace vehicles. The inclusion of calculation examples, diagrams, and software tool suggestions will greatly enhance the learning experience.

6.1 Introduction: The Art of Maintaining Control

Aerospace vehicles, unlike passive projectiles, require active control to maintain stability, achieve desired flight paths, and respond to changing environmental conditions. Control systems are the intricate networks of sensors, actuators, and algorithms that enable this precise manipulation of the vehicle's state. They translate pilot inputs (or autopilot commands) into actions that influence the aerodynamic forces and moments, thereby guiding the vehicle through the air or space.

In this chapter, we will explore the fundamental principles of control system design for aerospace vehicles. We will cover topics such as stability analysis, feedback control, PID control, state-space control, and various advanced control techniques. We will also discuss the challenges of designing robust control systems that can handle uncertainties and disturbances.

6.2 Open-Loop vs. Closed-Loop Control: A Fundamental Distinction

Control systems can be broadly classified into two categories: open-loop and closed-loop.

6.2.1 Open-Loop Control:

In an open-loop control system, the control action is independent of the system's output. The controller simply generates a control signal based on a pre-programmed input.

Open-loop control is simple and inexpensive but is susceptible to disturbances and uncertainties.

- **Example:**

A simple timer that turns on a light at a specific time is an example of open-loop control. The timer does not sense whether the light is actually on or off.

6.2.2 Closed-Loop Control (Feedback Control):

In a closed-loop control system, the control action is based on the system's output. The system's output is measured by sensors, and this information is fed back to the controller, which adjusts the control signal to minimize the error between the desired output (set point) and the actual output. Closed-loop control is more complex than open-loop control but offers greater accuracy and robustness.

- **Example:**

A thermostat is an example of closed-loop control. The thermostat senses the room temperature and adjusts the heating or cooling system to maintain the desired temperature.

6.3 The Anatomy of a Feedback Control System

A typical feedback control system consists of the following components:

- **Plant:** The system being controlled (e.g., the aerospace vehicle).

- **Sensors:** Devices that measure the output of the plant (e.g., accelerometers, gyroscopes, airspeed sensors, GPS).

- **Actuators:** Devices that exert control forces or moments on the plant (e.g., control surfaces, engines).

• **Controller:** The algorithm that generates the control signal based on the sensor measurements and the desired output (set point).

• **Summing Junction:** A point where the set point and the feedback signal are compared to generate an error signal.

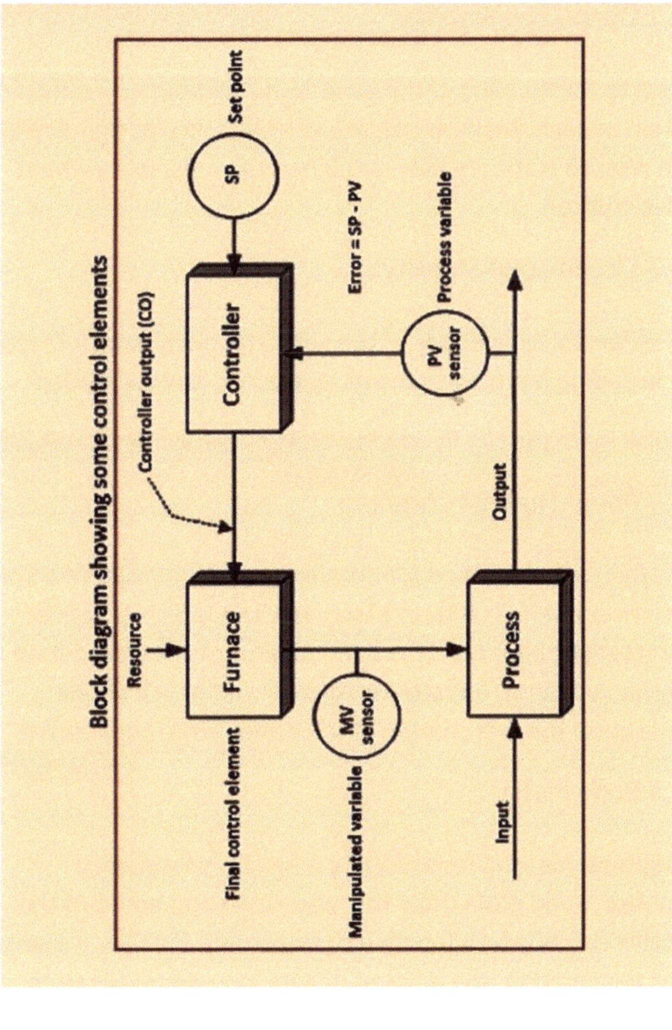

6.5 PID Control: The Workhorse of Feedback Systems

Proportional-Integral-Derivative (PID) control is one of the most widely used feedback control algorithms. It combines three control actions to generate a control signal based on the error between the desired set point and the measured output.

6.5.1 Proportional (P) Control:

The control signal is proportional to the error. A larger error produces a larger corrective action. Proportional control reduces the steady-state error but generally cannot eliminate it completely.

u(t)=Kp·e(t)u(t) = K_p \cdot e(t)u(t)=Kp·e(t)

Where K_p is the proportional gain, and e(t)e(t)e(t) is the error signal.

6.5.2 Integral (I) Control:

The control signal is proportional to the integral of the error over time. Integral control eliminates steady-state error by accumulating past errors. However, excessive integral action can lead to oscillations or instability.

u(t)=Ki∫e(t) dtu(t) = K_i \int e(t) \, dtu(t)=Ki∫e(t)dt

Where KiK is the integral gain.

6.5.3 Derivative (D) Control:

- The control signal is proportional to the rate of change of the error. Derivative control anticipates future errors, improving response time and damping. It can, however, amplify high-frequency noise.

u(t)=Kdde(t)dtu(t) = K_d \frac{d e(t)}{dt}u(t)=Kddtde(t)

Where KdK is the derivative gain.

The overall PID control signal combines these three actions:

u(t)=Kp·e(t)+Ki∫e(t) dt+Kdde(t)dtu(t) = K_p \cdot e(t) + K_i \int e(t) \, dt + K_d \frac{d e(t)}{dt}u(t)=Kp·e(t)+Ki∫e(t)dt+Kddtde(t)

PID gains (Kp,Ki,Kd) must be carefully tuned to achieve the desired performance. Common tuning methods include:

- **Trial and Error:** Adjusting gains iteratively based on system response.
- **Ziegler-Nichols Method:** A heuristic tuning method based on the system's open-loop response.
- **Optimization Algorithms:** Using numerical methods to find gains that minimize a performance criterion.

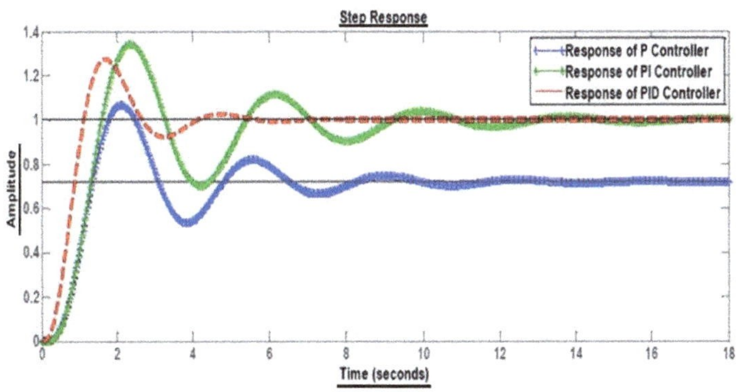

6.6 State-Space Control: A Modern Approach

State-space control is a modern control technique that provides a comprehensive and flexible framework for designing control systems. It models the system dynamics in terms of state variables, enabling advanced control strategies beyond classical methods.

6.6.1 State-Space Representation

A linear time-invariant (LTI) system can be described using the state-space representation:

State Equation:

- x˙=Ax+Bu\dot{x} = A x + B ux˙=Ax+Bu

Output Equation:

- y=Cx+Duy = C x + D uy=Cx+Du
- Where:
- x is the state vector, representing the internal state of the system.
- u is the input vector, representing control inputs.
- y is the output vector, representing measured outputs.
- A is the state matrix.
- B is the input matrix.
- C is the output matrix.
- D is the direct transmission matrix, often zero.

6.6.2 Controllability and Observability

Before designing a state-space controller, it is essential to check:

- **Controllability:** A system is controllable if it is possible to drive the system from any initial state to any desired final state in finite time using the control input.

- **Observability:** A system is observable if it is possible to determine the system's states from measurements of the outputs over a finite time interval.

6.6.3 Pole Placement

Pole placement allows the designer to arbitrarily assign the closed-loop poles of the system by designing a state feedback controller. The closed-loop poles determine stability and dynamic response.

The state feedback controller is:

u=−Kxu = -K xu=−Kx

Where K is the state feedback gain matrix, chosen to place the closed-loop poles at desired locations.

6.6.4 Linear Quadratic Regulator (LQR)

LQR is an **optimal control technique** that minimizes a cost function balancing performance and control effort:

J=∫(xTQx+uTRu) dtJ = \int (x^T Q x + u^T R u) \, dtJ=∫(xTQx+uTRu)dt

Where:

- Q is a positive semi-definite matrix penalizing deviations in the state.
- R is a positive definite matrix penalizing control effort.

The LQR controller has the form:

6.7.4 Model Predictive Control (MPC)

MPC uses a model of the system to predict future behavior over a finite time horizon and optimizes control inputs while considering constraints. It is well-suited for complex, multivariable systems with input/state limitations.

6.7.5 Fault-Tolerant Control

Fault-tolerant control systems maintain stability and performance even in the presence of faults in sensors, actuators, or other components. Methods often involve redundancy, reconfiguration, or adaptive compensation.

6.8 Control System Design for Specific Aerospace Vehicles

The design of a control system depends on the vehicle type and mission requirements.

6.8.1 Aircraft Control Systems

Aircraft typically use longitudinal (pitch), lateral (roll), and directional (yaw) control loops. These loops stabilize the aircraft, improve handling, and allow precise flight path control. Modern aircraft often include autopilot systems for tasks such as:

- Altitude hold
- Heading hold
- Navigation along a flight path

6.8.2 Spacecraft Attitude Control Systems

Spacecraft control systems maintain orientation in space. They commonly use:

- **Reaction Wheels**: Provide precise torque control for small adjustments.
- **Control Moment Gyros (CMGs)**: Generate large torques for rapid maneuvers.
- **Thrusters**: Used for coarse maneuvers or momentum dumping.

Precise pointing is critical for missions like Earth observation satellites, telescopes, and deep-space exploration.

6.8.3 Missile Guidance and Control Systems

Missile control systems guide the vehicle to its target using:

- **Inertial Navigation Systems (INS)**
- **GPS or radar inputs**

    These systems compute the missile's position and velocity and generate control commands to steer the missile along a desired trajectory.

6.9 Actuators and Sensors

Actuators and sensors form the physical interface between the control system and the aerospace vehicle. Proper selection is crucial to ensure accurate control and measurement.

Actuators

Common aerospace actuators include:

- **Hydraulic Actuators**: High force output, capable of moving large control surfaces quickly; heavier and more complex.
- **Electric Motors**: Lighter, simpler, and easier to integrate; limited force compared to hydraulics.
- **Control Surfaces**: Directly affect aerodynamic forces (e.g., ailerons, elevators, rudders).

Actuators must provide sufficient force, torque, and bandwidth to meet the dynamic requirements of the control system.

Sensors

Common aerospace sensors include:

- **Accelerometers**: Measure linear acceleration.
- **Gyroscopes**: Measure angular rates.
- **Airspeed Sensors**: Measure the vehicle's speed relative to the air.
- **GPS Receivers**: Provide position and velocity information.
- **Pressure Sensors**: Measure static or dynamic pressure for altitude and airspeed calculations.

Inertial Measurement Units (IMUs)

Combine accelerometers and gyroscopes to provide critical navigation and attitude information. Sensor selection must ensure accuracy, reliability, and appropriate sampling rates for control.

6.10 Calculation Examples

Example 1: PID Controller Design

- **System**: G(s)=1s+2G(s) = \frac{1}{s + 2}G(s)=s+21
- **Objective**: Achieve a closed-loop damping ratio ζ=0.707\zeta = 0.707ζ=0.707 and natural frequency ωn=5\omega_n = 5ωn=5 rad/s.

- **Method**: Use PID tuning techniques (e.g., Ziegler-Nichols, trial-and-error, or optimization algorithms) to determine KpK_pKp, KiK_iKi, and KdK_dKd.

Example 2: State Feedback Control

- **System**:
  A=[01−2−3],B=[01]A = \begin{bmatrix} 0 & 1 \\ -2 & -3 \end{bmatrix}, \quad B = \begin{bmatrix} 0 \\ 1 \end{bmatrix}A=[0−21−3],B=[01]

- **Objective**: Place closed-loop poles at -4 and -5.
- **Method**: Calculate the state feedback gain matrix K using pole placement techniques.

Example 3: LQR Control

- **System**: Same as Example 2.
- **Objective**: Design an LQR controller.
- **Parameters**: Q=[1001]Q = \begin{bmatrix} 1 & 0 \\ 0 & 1 \end{bmatrix}Q=[1001], R=[1]R = [1]R=[1]
- **Method**: Solve the Riccati equation to determine the optimal gain matrix K that minimizes the cost function:

J=∫(xTQx+uTRu) dtJ = \int (x^T Q x + u^T R u) \, dtJ=∫(xTQx+uTRu)dt

6.11 Software Tools for Control System Design

Several software tools are widely used in aerospace control system development:

- **MATLAB/Simulink**: Industry-standard platform for modeling, simulating, and designing control systems, with extensive toolboxes for PID, state-space, LQR, and MPC.

- **Python**: Libraries such as NumPy, SciPy, and control systems toolboxes provide flexible, open-source alternatives for control system analysis and simulation.
- **Modelica**: Object-oriented modeling language for simulating complex physical and mechatronic systems, suitable for integrating control and vehicle dynamics models.

6.12 Conclusion

Control systems are fundamental to the safe, stable, and efficient operation of aerospace vehicles. This chapter has provided a thorough overview of the principles and techniques used in control system design, including:

- **Stability Analysis** – ensuring predictable behavior in both open-loop and closed-loop systems.
- **Feedback Control** – using sensor measurements to correct errors and maintain desired states.
- **PID Control** – the classical workhorse of feedback systems, combining proportional, integral, and derivative actions.
- **State-Space Control** – a modern, flexible approach allowing pole placement, LQR design, and optimal control.
- **Advanced Control Techniques** – including adaptive, robust, nonlinear, and model predictive control.

By mastering these concepts, engineers can design robust, reliable, and high-performance control systems tailored to the demanding requirements of aerospace applications, from aircraft and spacecraft to missiles and UAVs.

6.13 Exercises

1. Explain the difference between open-loop and closed-loop control, and give examples of each in aerospace systems.

2. Define stability in the context of control systems. Why is it critical for aerospace vehicles?
3. Describe the three control actions in a PID controller and their individual effects on system response.
4. What is state-space control, and how does it differ from classical control techniques like PID or frequency-domain methods?
5. Research and compare the advantages and disadvantages of different control techniques:
   - PID control
   - State feedback control
   - LQR control
   - Adaptive control
   - Robust control
6. Design a PID controller for a simple DC motor speed control system. Include the tuning method and expected performance metrics.
7. Design a state feedback controller for a simple inverted pendulum system, ensuring closed-loop stability.
8. Explore the limitations of using a PID controller on an inherently unstable system and discuss how advanced control methods might improve performance.

Simulation Techniques and Validation: Testing Reality in the Virtual World

This chapter will delve into the art and science of simulation, covering numerical integration techniques, simulation software environments, validation methodologies, and best practices for building reliable and informative simulations.

7.1 Introduction: From Equations to Virtual Flight

In the previous chapters, we've constructed the mathematical foundation for modeling aerospace vehicle dynamics: coordinate systems, equations of motion, aerodynamic models, propulsion system models, and control system design. However, these equations, in their complexity, are often intractable for analytical solutions, especially for realistic scenarios. That's where simulation comes in.

Simulation provides a virtual environment where we can integrate these models, subject them to realistic conditions, and observe the resulting behavior of the aerospace vehicle. It allows us to test designs, explore flight envelopes, and validate control strategies without the risks and costs associated with physical prototypes or flight tests.

7.2 Numerical Integration: Solving the Equations of Motion

The core of any dynamic simulation is the numerical integration algorithm. These algorithms approximate the solution of differential equations (such as our equations of motion) by discretizing time and stepping forward from an initial state. Different algorithms have varying levels of accuracy, stability, and computational cost.

7.2.1 Euler Method:

- The simplest numerical integration method. It approximates the solution at the next time step using the current state and the derivative at the current time.

xn+1=xn+h·f(xn,tn)x_{n+1} = x_n + h \cdot f(x_n, t_n)xn+1=xn +h·f(xn,tn)

Where:

- xn+1x_{n+1}xn+1 is the state at the next time step (tn+1t_{n+1}tn+1).
- xnx is the state at the current time step (tnt_ntn).
- h is the time step size.
- f(xn,tn)f(x_n, t_n)f(xn,tn) is the derivative of the state at the current time step (x˙\dot{x}x˙).

The Euler method is easy to implement but has poor accuracy and stability, especially for stiff systems (systems with widely varying time constants). It is generally not recommended for aerospace vehicle simulations.

7.2.2 Runge-Kutta Methods:

- A family of more accurate and stable numerical integration methods. Runge-Kutta methods use multiple stages within each time step to estimate the derivative more accurately.
- **Fourth-Order Runge-Kutta (RK4):** A widely used Runge-Kutta method that provides a good balance between accuracy and computational cost. The RK4 method involves the following steps:

k1=h·f(xn,tn)k2=h·f(xn+k1/2,tn+h/2)k3=h·f(xn+k2/2,tn+h/2)k4= h·f(xn+k3,tn+h)xn+1=xn+k1+2k2+2k3+k46\begin{aligned} k_1 &= h \cdot f(x_n, t_n) \\ k_2 &= h \cdot f(x_n + k_1/2, t_n + h/2) \\ k_3 &= h \cdot f(x_n + k_2/2, t_n + h/2) \\ k_4 &= h \cdot f(x_n + k_3, t_n + h) \\ x_{n+1} &= x_n + \frac{k_1 + 2k_2 + 2k_3 + k_4}{6} \end{aligned}k1k2k3k4xn+1=h·f(xn,tn)=h·f(xn+k1/2,tn+h/2)=h·f(xn+k2/2,tn+h/2)=h·f(xn+k3,tn+h)=xn+6k1+2k2+2k3+k4

Where k1,k2,k3,k4k_1, k_2, k_3, k_4k1,k2,k3,k4 are intermediate derivative estimates.

RK4 is generally a good choice for aerospace vehicle simulations.

7.2.3 Variable Step-Size Methods:

- These methods automatically adjust the time step size based on the estimated error. They use smaller time steps when the solution is changing rapidly and larger time steps when the solution is changing slowly. Variable step-size methods can improve efficiency and accuracy, especially for systems with varying dynamics.
- Embedded Runge-Kutta methods provide error estimates that are used to adjust time step size.

7.2.4 Implicit Methods:

- Unlike explicit methods (like Euler and RK4), implicit methods use the solution at the next time step to estimate the derivative at the current time. Implicit methods are generally more stable than explicit methods, especially for stiff systems, but they require solving a system of nonlinear equations at each time step.
- Backward Euler is an example.

- **Define the Objectives:** Clearly define the objectives of the simulation. What questions are you trying to answer? What performance metrics are you interested in?
- **Develop the Models:** Develop mathematical models for the vehicle dynamics, aerodynamics, propulsion system, control system, and environmental effects.
- **Choose a Simulation Environment:** Select a simulation software environment that meets your needs.
- **Implement the Models:** Translate the mathematical models into computer code or block diagrams within the chosen simulation environment.
- **Integrate the Models:** Connect the different models together to create a complete simulation.
- **Configure the Simulation:** Set the initial conditions, simulation parameters, and simulation time.
- **Run the Simulation:** Execute the simulation and collect the results.
- **Analyze the Results:** Analyze the simulation results and draw conclusions.
- **Validate the Simulation:** Compare the simulation results with experimental data or other reliable sources to ensure that the simulation is accurate.
- **Iterate and Refine:** Modify the models or simulation parameters based on the validation results and repeat the process.

7.5 Simulation Validation: Ensuring Accuracy and Trust

Simulation validation is the process of determining the degree to which a simulation is an accurate representation of the real world from the perspective of the intended uses of the simulation. It is a critical step in building confidence in the simulation results.

7.5.1 Validation Data:

Validation requires having reliable data from experiments, flight tests, or other validated simulations to compare against.

7.5.2 Validation Metrics:

Define quantitative metrics for comparing the simulation results with the validation data. Common metrics include:

- **Root Mean Square Error (RMSE):** A measure of the average difference between the simulation results and the validation data.
- RMSE=Σ(ysim−yexp)2N\text{RMSE}=\sqrt{\frac{\Sigma(y_\text{sim} -y_\text{exp})^2}{N}}RMSE=NΣ(ysim−yexp)2

Where:

.1 ysimy_\text{sim}ysim is the simulation result.
.2 yexpy_\text{exp}yexp is the experimental data.
.3 N is the number of data points.

- **Mean Absolute Error (MAE):** A measure of the average absolute difference between the simulation results and the validation data.

MAE=Σ|ysim−yexp|N\text{MAE} = \frac{\Sigma |y_\text{sim} - y_\text{exp}|}{N}MAE=NΣ|ysim−yexp|

- **Correlation Coefficient (R^2):** A measure of the linear relationship between the simulation results and the validation data. An R2R^2R2 value of 1 indicates a perfect correlation.

7.5.3 Uncertainty Quantification:

Account for the uncertainties in the models, simulation parameters, and validation data. Perform sensitivity analyses to determine which parameters have the greatest impact on the simulation results.

Sensitivity=Change in OutputChange in Input\text{Sensitivity} =\frac{\text{Change in Output}}{\text{Change in Input}}Sensitivity=Change in InputChange in Output.

7.5.4 Calibration:

Adjust the model parameters to improve the agreement between the simulation results and the validation data. This process is known as calibration or parameter estimation. Note that parameters should only be adjusted within physically realistic ranges.

7.6 Hardware-in-the-Loop (HIL) Simulation

Hardware-in-the-Loop (HIL) simulation is a technique that integrates physical hardware components into the simulation loop. This allows for more realistic testing of the hardware and software interactions.

7.6.1 Benefits of HIL Simulation:

- **Increased Fidelity:** HIL simulation provides a more realistic testing environment than purely software-based simulation.

- **Early Detection of Errors:** HIL simulation can help detect errors in the hardware or software design early in the development process.
- **Reduced Risk:** HIL simulation can reduce the risk of damaging expensive hardware during testing.
- Provides real-time performance.

7.6.2 HIL Simulation Setup:

A typical HIL simulation setup includes:

- **Real-Time Computer:** A computer that runs the simulation in real time.
- **Input/output (I/O) Interfaces:** Interfaces that connect the real-time computer to the hardware components.
- **Hardware Components:** The physical hardware components being tested (e.g., flight control computer, actuators, sensors).
- **Software Models:** Software models that represent the remaining components of the system.

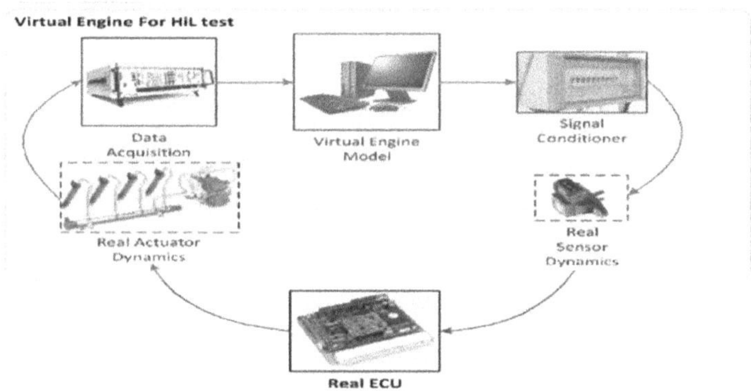

7.9 Best Practices for Simulation

- **Document Everything:** Document the models, assumptions, simulation parameters, and validation results clearly and thoroughly.
- **Use Version Control:** Use a version control system (e.g., Git) to track changes to the simulation code and models.
- **Test Your Code:** Write unit tests to verify the correctness of the simulation code.
- **Use Modular Design:** Design the simulation in a modular way, with well-defined interfaces between the different components.
- **Validate Incrementally:** Validate the simulation incrementally, starting with the simplest components and gradually adding complexity.
- **Perform a Failure Modes and Effects Analysis (FMEA):** Identify potential simulation failures and their impact on the results.

7.10 Example Calculations

- **Example 1: RK4 Integration**

Assume a simple differential equation:

x˙=−2x\dot{x} = -2xx˙=−2x

and a step size of h=0.1h = 0.1h=0.1 seconds, and an initial condition of x(0)=1x(0) = 1x(0)=1. Let's calculate the next value of x using RK4.

1. k1=hf(xn,tn)=0.1(−2⋅1)=−0.2k_1 = h f(x_n, t_n) = 0.1 (-2 \cdot 1) = -0.2k1=hf(xn,tn)=0.1(−2⋅1)=−0.2
2. k2=hf(xn+k1/2,tn+h/2)=0.1(−2(1−0.22))=−0.18k_2 = h f(x_n + k_1/2, t_n + h/2) = 0.1 \left(-2\left(1 -

\frac{0.2}{2}\right)\right) = -0.18k2=hf(xn+k1/2,tn +h/2)=0.1(−2(1−20.2))=−0.18

3. k3=hf(xn+k2/2,tn+h/2)=0.1(−2(1−0.182))=−0.182k_3 = h f(x_n + k_2/2, t_n + h/2) = 0.1 \left(-2\left(1 - \frac{0.18}{2}\right)\right) = -0.182k3=hf(xn+k2/2,tn +h/2)=0.1(−2(1−20.18))=−0.182
4. k4=hf(xn+k3,tn+h)=0.1(−2(1−0.182))=−0.1636k_4 = h f(x_n + k_3, t_n + h) = 0.1 \left(-2(1 - 0.182)\right) = -0.1636k4 =hf(xn+k3,tn+h)=0.1(−2(1−0.182))=−0.1636

Then,

xn+1=xn+k1+2k2+2k3+k46x_{n+1} = x_n + \frac{k_1 + 2k_2 + 2k_3 + k_4}{6}xn+1=xn+6k1+2k2+2k3+k4 xn+1=1+−0.2+2(−0.18)+2(−0.182)+(−0.1636)6x_{n+1} = 1 + \frac{-0.2 + 2(-0.18) + 2(-0.182) + (-0.1636)}{6}xn+1 =1+6−0.2+2(−0.18)+2(−0.182)+(−0.1636) xn+1=0.8187x_{n+1} = 0.8187xn+1=0.8187

- **Example 2: RMSE Calculation**

A simulation predicted velocities of [100,102,105][100, 102, 105][100,102,105] m/s at three time points, while the measured velocities were [98,103,106][98, 103, 106][98,103,106] m/s. Calculate the RMSE.

RMSE=(100−98)2+(102−103)2+(105−106)23\text{RMSE} = \sqrt{\frac{(100 - 98)^2 + (102 - 103)^2 + (105 - 106)^2}{3}}RMSE=3(100−98)2+(102−103)2+(105−106)2 RMSE=4+1+13=1.414 m/s\text{RMSE} = \sqrt{\frac{4 + 1 + 1}{3}} = 1.414 \text{ m/s}RMSE=34+1+1=1.414 m/s

7.11 Conclusion

Simulation is an indispensable tool for designing, analyzing, and testing aerospace vehicles. This chapter has provided a comprehensive overview of the key techniques and methodologies used in simulation, covering numerical integration, simulation software environments, validation, HIL simulation, and best practices. By mastering these concepts, engineers can build reliable and informative simulations that provide valuable insights into the behavior of aerospace vehicles and enable them to make informed design decisions.

7.12 Exercises

1. Explain the importance of numerical integration in aerospace vehicle simulation.
2. Compare and contrast different numerical integration methods (Euler, RK4, variable step-size).
3. What is simulation validation, and why is it important?
4. Describe the steps involved in building a reliable aerospace vehicle simulation.
5. What is HIL simulation, and what are its benefits?
6. Write a Python function that implements the RK4 method for a simple differential equation.
7. Design a simulation to analyze the effect of wind gusts on the stability of an aircraft.

Case Studies in Aerospace Vehicle Dynamics Simulation: From Theory to Practice

8.2.5 Actuators:

The rotor speeds of the four motors, which are controlled by the PID controllers.

8.2.6 Sensors:

• **Inertial Measurement Unit (IMU):** Measures angular rates (p, q, r) and linear accelerations (ax,ay,aza_x, a_y, a_zax,ay,az) in the body frame.

• **Altimeter:** Measures altitude above ground level.

8.2.7 Simulation Setup:

• Use MATLAB/Simulink or Python to implement the models and equations of motion.

• Choose a numerical integration method (e.g., RK4) with a fixed time step.

• Set the initial conditions for the quadcopter's position, velocity, and attitude.

• Define the control inputs (throttle, roll, pitch, yaw) as functions of time.

8.2.8 Controller Design:

The quadcopter controller uses four cascaded PID loops:

1. Altitude hold.
2. Roll angle and pitch angle control.
3. Yaw angle control.

The outputs of these PID loops command the desired motor speeds.

8.2.9 Example Calculation: Thrust Generation

Assume each motor produces a thrust T proportional to the square of its angular velocity ω\omegaω:

T=kTω2T = k_T \omega^2T=kTω2

Where kTk_TkT is the thrust coefficient.

If kT=1×10−6k_T = 1 \times 10^{-6}kT=1×10−6 N/(rad/s)2^22 and the motor is spinning at 1000 rad/s, the thrust produced by the motor is:

T=1×10−6⋅(1000)2=1 NT = 1 \times 10^{-6} \cdot (1000)^2 = 1 \text{ N}T=1×10−6⋅(1000)2=1 N

The total thrust of the quad copter is the sum of the thrusts produced by all four motors. The total force vector in the body frame must then be transformed to the inertial frame using the rotation matrix.

8.2.10 Simulation Results:

• Plot the quad copter's position, velocity, and attitude as functions of time.

• Analyze the quad copter's response to different control inputs.

• Evaluate the performance of the PID controllers.

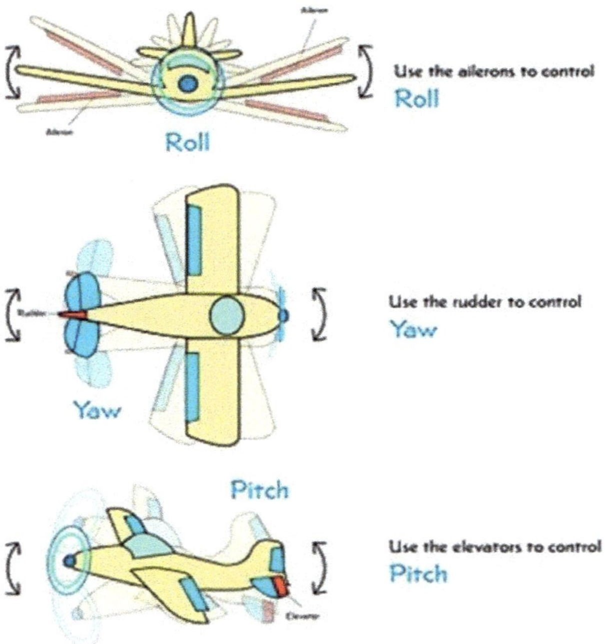
Roll
Use the ailerons to control
Roll
Yaw
Use the rudder to control
Yaw
Pitch
Use the elevators to control
Pitch

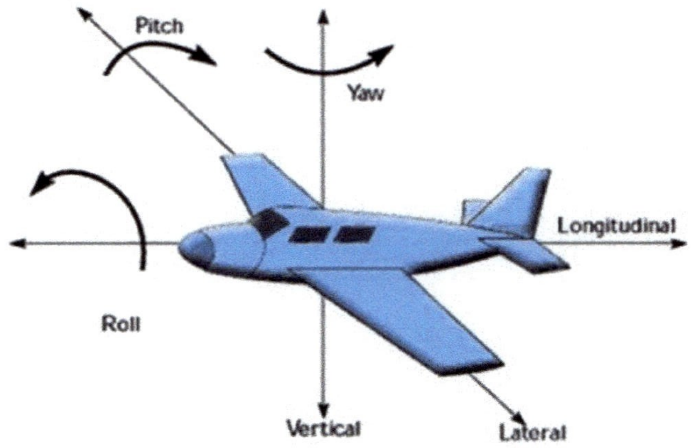
Pitch
Yaw
Roll
Longitudinal
Vertical
Lateral

8.2.11 Validation:

• Compare the simulation results with experimental data from flight tests of a real quad copter.

• Tune the model parameters and PID gains to improve the agreement between the simulation results and the experimental data.

8.2.12 Enhancements:

• Model sensor noise and biases.

• Add a wind disturbance model.

• Create an automated path-following controller.

8.3 Case Study 2: Simulating the Trajectory of a Rocket Launch

8.3.1 Objectives:

• Develop a dynamic model of a rocket launching from Earth.

• Simulate its trajectory from liftoff to orbit insertion.

• Analyze the effects of gravity, atmospheric drag, and thrust on the rocket's trajectory.

• Design a simple guidance system to steer the rocket to a desired orbit.

8.3.2 Models:

• Rigid Body Dynamics: Model the rocket as a rigid body with six degrees of freedom.

8.3.9 Example Calculation: Thrust Required

To achieve a target orbital velocity, the rocket will use the Tsiolkovsky rocket equation to define the velocity change needed for the burn.

$\Delta V = Isp * g_0 * \ln(Mi / Mf)$
Where:

- ΔV is the change in velocity,
- Isp is the rocket's specific impulse,
- g_0 is the gravitational acceleration (9.81 m/s^2),
- Mi is the initial mass (propellant + dry mass),
- Mf is the final mass (dry mass).

To compute the thrust T required for burn time t:
T = mDot * Ve

8.3.10 Simulation Results:

• Plot the rocket's trajectory (position, velocity, and altitude) as functions of time.

• Analyze the effects of gravity, atmospheric drag, and thrust on the rocket's trajectory.

• Evaluate the performance of the guidance system.

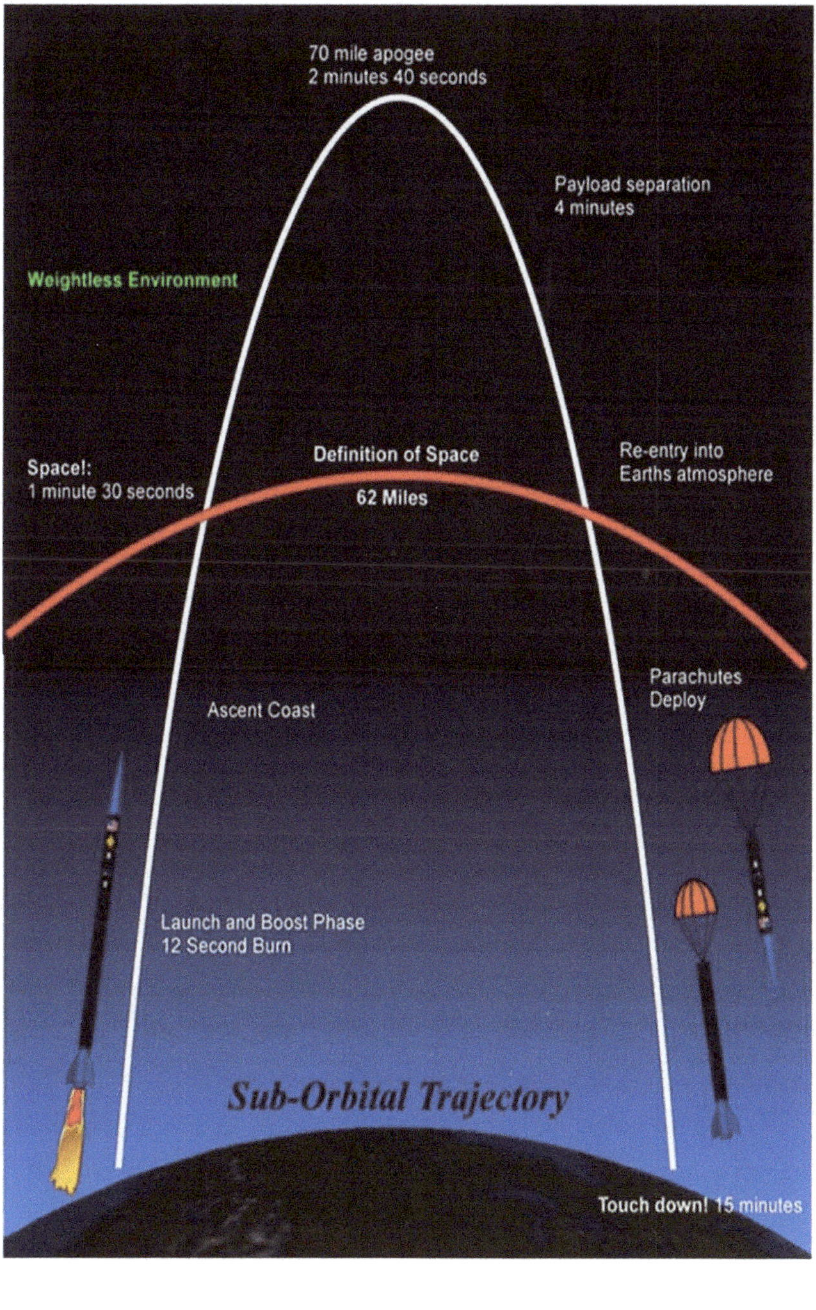
70 mile apogee
2 minutes 40 seconds
Payload separation
4 minutes
Weightless Environment
Space!:
1 minute 30 seconds
Definition of Space
62 Miles
Re-entry into
Earths atmosphere
Parachutes
Deploy
Ascent Coast
Launch and Boost Phase
12 Second Burn
Sub-Orbital Trajectory
Touch down! 15 minutes

8.3.11 Validation:

• Compare the simulation results with real-world rocket launch data.

• Tune the model parameters and guidance system parameters to improve the agreement between the simulation results and the real-world data.

8.3.12 Enhancements:

• Multi-stage rockets

• Fin modeling for aerodynamic stability

• Wind modeling

8.4 Case Study 3: Simulating the Approach and Landing of a Space Shuttle (or Reusable Launch Vehicle)

8.4.1 Objectives:

• Develop a dynamic model of a Space Shuttle (or Reusable Launch Vehicle) during its approach and landing phase.

• Simulate its descent from high altitude to touchdown on the runway.

• Analyze the effects of aerodynamic forces, control surface deflections, and wind gusts on the vehicle's trajectory.

• Design a control system to enable a safe and precise landing.

8.4.2 Models:

• Rigid Body Dynamics: Model the Shuttle as a rigid body with six degrees of freedom.

• Aerodynamic Model: Use aerodynamic look-up tables or empirical equations to model the aerodynamic forces and moments as functions of angle of attack, sideslip angle, Mach number, and control surface deflections.

• Atmospheric Model: Model the atmospheric density, temperature, and wind using a standard atmospheric model.

• Control System: Implement a control system to automatically control the Shuttle's attitude and glide path.

8.4.3 Coordinate Frames:

• Inertial Frame (I-Frame): Local tangent frame at the landing site.

• Body Frame (B-Frame): Centered at the Shuttle's center of gravity, with axes aligned with the Shuttle's principal axes of inertia.

• Wind Frame (W-Frame): Aligned with the relative wind velocity vector.

8.4.4 Equations of Motion:

• Same as outlined in previous sections.

8.4.5 Actuators:

- Control surface deflections (elevons, rudder, speed brake).

8.4.6 Sensors:

• Inertial Measurement Unit (IMU): Measures angular rates and linear acceleration.

• Airspeed Sensor: Measures airspeed.

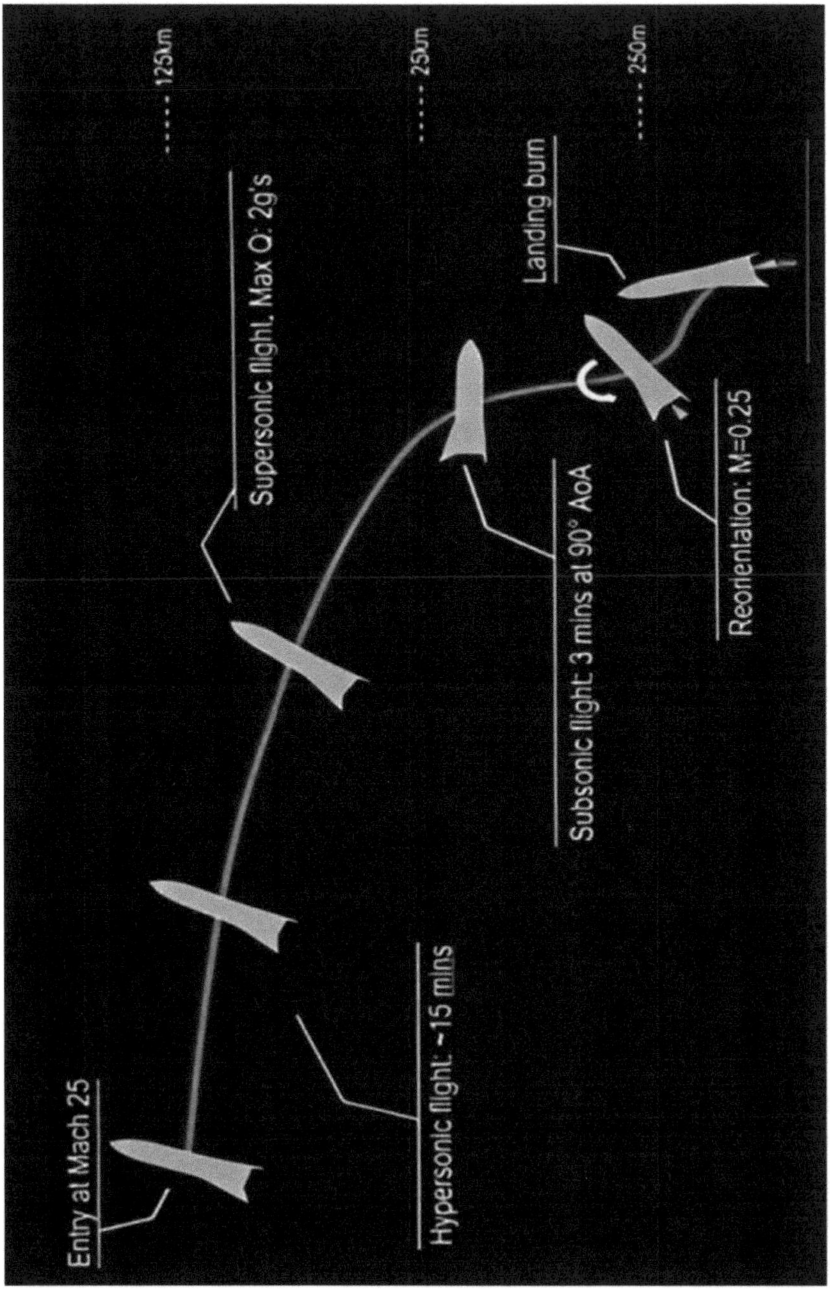
Entry at Mach 25
Hypersonic flight: ~15 mins
Supersonic flight. Max Q: 2g's
Subsonic flight: 3 mins at 90° AoA
Reorientation: M=0.25
Landing burn
125km
25km
250m

8.4.11 Validation:

• Compare the simulation results with flight test data from real Space Shuttle landings (or flight tests of a similar vehicle)

• Tune the model parameters and control system parameters to improve the agreement between the simulation results and the flight test data.

8.4.12 Enhancements:

• Landing Gear Modeling

• Runway Model

• Thrust Reversers

8.5 General Enhancements for All Case Studies:

• Incorporate a pilot-in-the-loop simulation.

• Develop a Fault Detection, Isolation, and Recovery (FDIR) system.

• Explore the use of machine learning to improve simulation fidelity and efficiency.

• Sensitivity analysis.

8.6 Conclusion: The Power of Simulation

These case studies demonstrate the power and versatility of aerospace vehicle dynamics simulation. By combining theoretical knowledge with practical implementation, engineers can gain valuable insights into the behavior of complex systems, design effective control strategies, and validate their designs before committing to expensive and risky physical prototypes or flight tests. As computational power continues to increase and simulation tools become more sophisticated, the role of simulation in aerospace engineering will only continue to grow.

8.7 Exercises

1. For each of the case studies presented in this chapter, identify the key assumptions and simplifications that were made in the models. How might these assumptions affect the accuracy of the simulation results?
2. Choose one of the case studies and develop a more detailed model for one of the components (e.g., the engine model, the aerodynamic model, or the control system model).
3. For each of the case studies, suggest additional validation data sources that could be used to improve the confidence in the simulation results.
4. Research a real-world aerospace accident or incident and analyze how simulation could have been used to prevent the event.
5. Develop a simulation of a simple aircraft performing a coordinated turn. Analyze the effects of bank angle and airspeed on the turn radius and load factor.

Emerging Trends and Future Directions: Charting the Course of Aerospace Simulation

9.1 Introduction: A Glimpse into the Future of Virtual Flight

The field of aerospace vehicle dynamics simulation is constantly evolving, driven by advancements in computing power, sensor technology, and control algorithms. As we stand on the cusp of new aerospace innovations, from autonomous drones to hypersonic aircraft and interplanetary missions, the need for accurate and efficient simulation tools has never been greater.

This chapter will explore some of the most exciting emerging trends and future directions in the field, offering a glimpse into the capabilities that will shape the next generation of aerospace vehicle design, testing, and operation. We will explore topics ranging from machine learning and digital twins to advanced visualization and the potential of quantum computing.

9.2 Machine Learning and Artificial Intelligence: Transforming Simulation

Machine learning (ML) and artificial intelligence (AI) are rapidly transforming various fields, and aerospace vehicle dynamics simulation is no exception. These technologies offer the potential to improve the accuracy, efficiency, and robustness of simulations in several ways:

9.2.1 Surrogate Modeling:

High-fidelity simulations, such as Computational Fluid Dynamics (CFD), can be computationally expensive. Machine learning can be used to create surrogate models that approximate the behavior of these high-fidelity simulations with much lower computational cost.

- **Example:** Train a neural network on CFD data to predict aerodynamic coefficients as a function of angle of attack, sideslip angle, Mach number, and control surface deflections. This neural network can then be used in real-time flight simulations or control system design.
- **Techniques:** Neural Networks (NNs), Support Vector Machines (SVMs), Gaussian Process Regression (GPR).
- **Benefits:** Significant reduction in computational time, enabling real-time simulation and optimization.

9.2.2 System Identification:

Machine learning can be used to identify the parameters of a system model from experimental or simulation data. This is particularly useful for systems with complex or uncertain dynamics.

- **Example:** Use a recursive least squares (RLS) algorithm or a Kalman filter to estimate the aerodynamic derivatives of an aircraft from flight-test data.
- **Techniques:** Recursive Least Squares, Kalman Filters, Neural Networks.
- **Benefits:** Improved model accuracy, adaptive control system design.

9.2.3 Control System Design:

Machine learning can be used to design and tune control systems for aerospace vehicles. Reinforcement learning algorithms can train controllers to achieve desired performance objectives in complex and uncertain environments.

- **Example:** Use a Deep Q-Network (DQN) to train a controller to land a drone autonomously in the presence of wind gusts.
- **Techniques:** Reinforcement Learning (RL), Deep Learning, Evolutionary Algorithms.
- **Benefits:** Improved control performance, robustness to uncertainties, autonomous operation.

9.2.4 Anomaly Detection:

Machine learning algorithms can analyze simulation data to detect anomalies or deviations from expected behavior, potentially indicating design flaws or unexpected system interactions.

9.2.5 Computational Example: Neural Network Surrogate Model

Consider using a neural network to create a surrogate model of an airfoil's lift coefficient (Cl) based on angle of attack (alpha).

1. **Data Generation:** Use CFD simulations to generate a dataset of (alpha, Cl) pairs.
2. **Network Architecture:** Define a neural network with an input layer (alpha), a hidden layer, and an output layer (Cl).
3. **Training:** Train the neural network using the CFD dataset with a suitable optimization algorithm (e.g., Adam).
4. **Validation:** Validate the surrogate model by comparing its predictions with CFD data.

Python Code (Using TensorFlow/Keras)

```
import numpy as np
```

```
import tensorflow as tf
from tensorflow import keras

# Generate synthetic CFD data (replace with your actual CFD data)
alpha = np.linspace(-10, 10, 100)  # Angle of attack in degrees
Cl = 0.2 * alpha + 0.01 * alpha**2 + np.random.normal(0, 0.05, 100)  # Simulate Cl

# Define the neural network model
model = keras.Sequential([
    keras.layers.Dense(64, activation='relu', input_shape=[1]),
    keras.layers.Dense(64, activation='relu'),
    keras.layers.Dense(1)
])

# Compile the model
optimizer = tf.keras.optimizers.Adam(0.001)
model.compile(loss='mse', optimizer=optimizer, metrics=['mae', 'mse'])

# Train the model
history = model.fit(alpha, Cl, epochs=100, validation_split=0.2, verbose=0)

# Evaluate the model
loss, mae, mse = model.evaluate(alpha, Cl, verbose=0)
print(f"Mean Absolute Error: {mae}")
print(f"Mean Squared Error: {mse}")
```

This simplified code illustrates how to build and train a neural network to approximate the relationship between angle of attack and lift coefficient.

9.3 Digital Twins: A Virtual Mirror of Reality

A digital twin is a virtual representation of a physical aerospace vehicle or system that is continuously updated with real-time data from sensors. Digital twins can be used for:

9.3.1 Performance Monitoring:

Monitor the health and performance of the vehicle in real time and detect anomalies or potential failures.

9.3.2 Predictive Maintenance:

Predict when maintenance is required based on the vehicle's operating conditions and performance data.

9.3.3 Simulation and Optimization:

Use the digital twin to simulate the vehicle's behavior under different operating conditions and optimize its performance.

9.3.4 Training:

Use the digital twin as a training simulator for pilots and maintenance personnel.

9.3.5 Design Iteration:

Use real-world performance data from the digital twin to inform and refine future designs.

9.3.6 Enhanced Control:

Integrate the digital twin with the control system to provide more accurate and robust control.

DIGITAL TWIN

9.3.7 Real-World Example: Engine Health Monitoring

A digital twin of an aircraft engine could monitor parameters like temperature, pressure, vibration, and oil quality. By analyzing this data, the digital twin can detect anomalies, predict maintenance needs, and optimize engine performance. If the twin notices higher-than-expected turbine blade temperatures, it might flag a potential issue requiring maintenance.

9.4 Advanced Visualization and Augmented Reality: Immersive Simulation

Advanced visualization techniques, such as virtual reality (VR) and augmented reality (AR), are enhancing the way engineers interact with simulations.

9.4.1 Virtual Reality (VR):

VR allows engineers to immerse themselves in a virtual environment that represents the aerospace vehicle or its operating environment. This can be used for:

- **Design Review:** Review the design of the vehicle in a realistic 3D environment.
- **Training:** Train pilots and maintenance personnel in a safe and immersive environment.
- **Mission Planning:** Plan missions and rehearse procedures in a virtual environment.
- **Human-in-the-Loop Simulation:** Integrate human operators into the simulation loop to evaluate the vehicle's handling qualities and control system performance.

9.4.2 Augmented Reality (AR):

AR overlays virtual information onto the real world. This can be used for:

- **Maintenance and Repair:** Provide technicians with real-time guidance and information during maintenance and repair procedures.
- **Flight Testing:** Overlay flight test data onto the pilot's view during flight testing.
- **Pilot Assistance:** Overlay navigation information and other flight data onto the pilot's view during flight.

9.4.3 Example: VR Cockpit Simulation

A VR cockpit simulation allows pilots to experience realistic flight conditions and practice maneuvers in a safe and controlled environment. This can be used to train pilots for emergency situations, evaluate new cockpit designs, and assess the impact of different control system parameters.

9.5 High-Fidelity Modeling: Capturing Complex Physics

While simplified models are often used for real-time simulation, high-fidelity models are essential for capturing complex physical phenomena that can significantly affect the behavior of aerospace vehicles.

9.5.1 Computational Fluid Dynamics (CFD):

Solve the Navier–Stokes equations to simulate the airflow around aerospace vehicles with high accuracy. This is particularly important for analyzing transonic and supersonic flows, stall phenomena, and complex aerodynamic interactions.

9.5.2 Finite Element Analysis (FEA):

Simulate the structural behavior of aerospace vehicles under different loading conditions. This is important for analyzing stress, strain, and deformation, and for designing lightweight and robust structures.

9.5.3 Multi-Physics Simulation:

Combine different simulation techniques (e.g., CFD, FEA, thermal analysis) to capture the complex interactions between different physical phenomena.

This is particularly important for hypersonic vehicles, where aerodynamic heating, structural deformation, and propulsion system performance are strongly coupled.

9.5.4 Reduced Order Modeling (ROM):

Reduces the complexity of a high-fidelity model to lower computational cost while preserving the important characteristics of the high-fidelity model.

9.5.5 Example: Hypersonic Vehicle Simulation:

Simulating the flight of a hypersonic vehicle requires capturing the complex interactions between aerodynamics, propulsion, and thermal effects. CFD simulations can be used to model the airflow around the vehicle, accounting for shock waves, boundary-layer transition, and aerodynamic heating. FEA simulations can be used to model the structural deformation of the vehicle due to aerodynamic loads and thermal stresses.

9.6 Autonomous Systems and Swarm Simulation: A New Era of Flight

Autonomous systems, such as drones and self-flying aircraft, are becoming increasingly prevalent in the aerospace industry.

Simulating the behavior of these autonomous systems is essential for ensuring their safety and reliability.

9.6.1 Multi-Agent Simulation:

Simulate the interactions between multiple autonomous vehicles in a shared environment. This is particularly important for analyzing swarm behavior, collision avoidance, and task allocation.

9.6.2 Sensor Fusion:

Simulate the integration of data from multiple sensors to improve the accuracy and robustness of the autonomous system's perception of its environment.

9.6.3 Decision-Making Algorithms:

Simulate the decision-making algorithms used by the autonomous system to plan its path and execute its tasks.

9.6.4 Example: Drone Delivery Swarm

Simulating a drone delivery swarm requires modeling the individual drones, their communication network, and their environment. The simulation can be used to analyze the swarm's ability to deliver packages efficiently and reliably, even in the presence of obstacles, wind gusts, and other disturbances.

9.6.5 Swarm Optimization:

Algorithms such as Particle Swarm Optimization (PSO) can be integrated into simulations to optimize the behavior of vehicle swarms. This could involve optimizing the formation, path planning, or task allocation of the swarm to achieve specific objectives.

9.6.6 Calculation Example: Particle Swarm Optimization

Imagine optimizing the trajectory of a UAV swarm to survey an area. Each UAV represents a particle, and its trajectory is defined by a set of parameters (e.g., waypoints, speed).

1. **Initialization:** Initialize the positions (trajectories) and velocities of the particles randomly.
2. **Fitness Evaluation:** Evaluate the fitness of each trajectory based on a cost function that considers factors like coverage area, flight time, and collision avoidance.
3. **Update Personal Best:** For each particle, update its personal best position (trajectory) if its current position is better than its previous best.
4. **Update Global Best:** Identify the particle with the best fitness in the entire swarm and update the global best position.
5. **Update Velocity and Position:** Update the velocity and position of each particle based on the following equations:

vit+1=w·vit+c1·r1·(pi−xit)+c2·r2·(g−xit)v_i^{t+1} = w \cdot v_i^t + c_1 \cdot r_1 \cdot (p_i - x_i^t) + c_2 \cdot r_2 \cdot (g - x_i^t)vit+1=w·vit+c1·r1·(pi−xit)+c2·r2·(g−xit)
xit+1=xit+vit+1x_i^{t+1} = x_i^t + v_i^{t+1}xit+1=xit+vit+1

Where:

- vitv_i^tvit is the velocity of particle i at iteration t.
- xitx_i^txit is the position of particle i at iteration t.
- w is the inertia weight.
- c1c and c2c are acceleration coefficients.
- r1r and r2r are random numbers between 0 and 1.
- pip is the personal best position of particle i.
- G is the global best position.

6. **Repeat Steps 2–5** until convergence or a maximum number of iterations is reached.

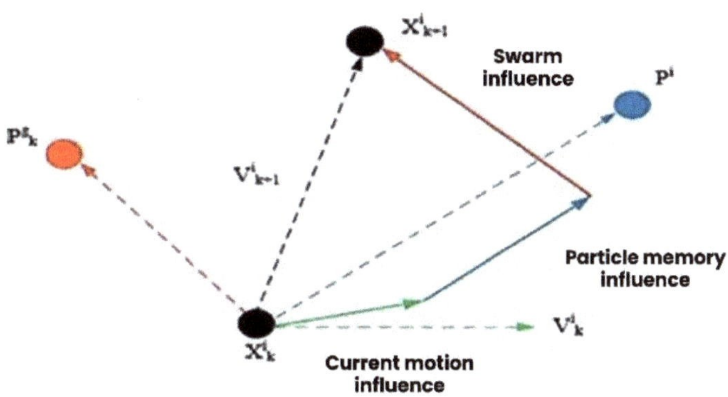

9.7 Quantum Computing: A Revolutionary Leap in Computational Power

Quantum computing is an emerging technology that has the potential to revolutionize many fields, including aerospace vehicle dynamics simulation. Quantum computers use quantum bits (qubits) to perform computations, which can allow them to solve certain types of problems much faster than classical computers.

9.7.1 Applications in Aerospace Simulation:

- **CFD:** Quantum algorithms could be used to solve the Navier-Stokes equations much faster than classical algorithms, enabling high-fidelity simulations of complex aerodynamic flows in real time.
- **FEA:** Quantum algorithms could be used to solve large-scale FEA problems much faster than classical algorithms, enabling the design of lightweight and robust aerospace structures.
- **Optimization:** Quantum algorithms could be used to optimize the design of aerospace vehicles and control systems.

9.7.2 Challenges

- Quantum computing is still in its early stages of development.
- Quantum computers are expensive and difficult to build and maintain.
- Quantum algorithms are not yet available for all types of problems.

9.7.3 Example: Quantum-Accelerated CFD

While still theoretical, quantum algorithms could potentially be used to solve the Navier-Stokes equations, enabling real-time CFD simulations. This would revolutionize the design and testing of aerospace vehicles.

9.8 Ethical Considerations

As simulations become more powerful and influence decision-making, ethical considerations become paramount. Algorithmic bias, data privacy, and the potential for misuse of simulations all require careful attention. Engineers must strive to create simulations that are fair, transparent, and used responsibly.

9.9 Validation in the Age of Machine Learning

Traditional validation techniques may not be sufficient for machine-learning-based simulations. New validation methods are needed to assess the reliability and robustness of these models, including adversarial testing and explainable AI techniques.

9.10 Conclusion: Embracing the Future of Simulation

The field of aerospace vehicle dynamics simulation is poised for a period of rapid innovation, driven by the convergence of machine learning, digital twins, advanced visualization, and quantum computing. By embracing these emerging trends and developing new simulation techniques, engineers can unlock new possibilities for designing, testing, and operating aerospace vehicles, enabling them to explore new frontiers in air and space.

9.11 Exercises

1. Research and write a report on the current state of quantum computing and its potential applications in aerospace engineering.
2. Develop a machine-learning-based surrogate model for a simple aerodynamic system.
3. Design a digital twin for a UAV delivery system.
4. Explore the use of virtual reality for pilot training or maintenance and repair procedures.
5. Discuss the ethical considerations of using machine learning and AI in aerospace vehicle dynamics simulation.
6. Examine the potential security risks of digital twins.
7. What impact would more efficient CFD have on airplane design?
8. What data privacy concerns exist when using flight data to train a machine learning model?

Real-Time Simulation and Hardware-in-the-Loop Testing: Bringing the Virtual and Physical Worlds Together

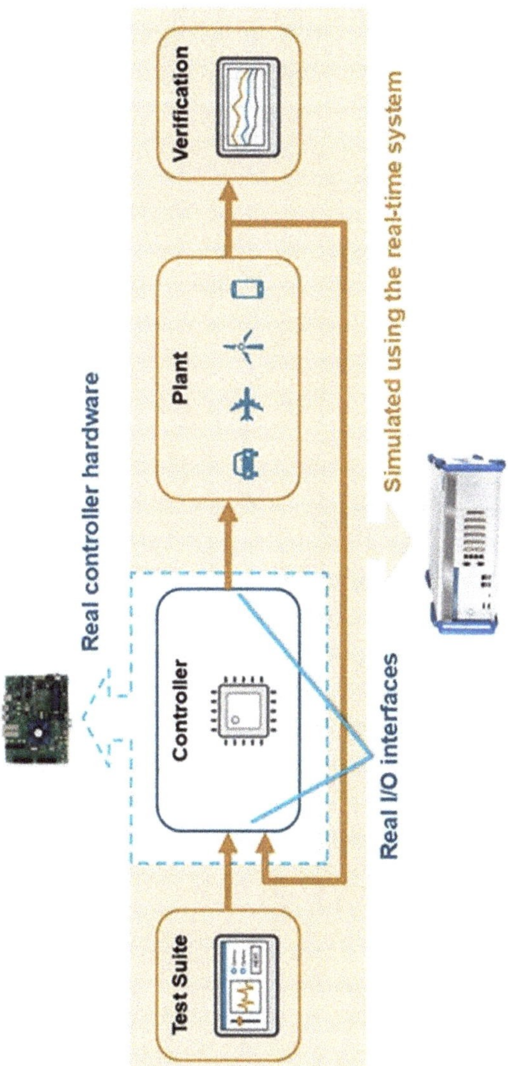

10.1 Introduction: Stepping into the Real-Time Realm

In previous chapters, we have discussed various aspects of aerospace vehicle dynamics simulation, from mathematical modeling to numerical integration and control system design. We have primarily focused on offline simulations, where the simulation runs at a pace independent of real time. However, many aerospace applications require simulations that operate in real time, meaning that the simulation time progresses at the same rate as actual time.

Real-time simulation is crucial for several reasons: it allows for human-in-the-loop testing, enables the integration of physical hardware components into the simulation loop (Hardware-in-the-Loop, or HIL), and facilitates the development and testing of real-time control systems.

This chapter will explore the principles and techniques of real-time simulation and HIL testing for aerospace systems. We will cover topics such as real-time operating systems, I/O interfaces, hardware modeling, and validation methodologies. We will also discuss the challenges of achieving real-time performance and the benefits of using HIL testing to improve the reliability and safety of aerospace systems.

10.2 The Need for Speed: What is Real-Time Simulation?

Real-time simulation is defined as a simulation where the simulation time progresses at the same rate as actual time. This means that the simulation must be able to execute all of its calculations within a fixed time step, ensuring that the simulation output is synchronized with the real world.

10.2.1 Importance of Determinism:

Achieving real-time performance requires determinism, meaning that the simulation must produce the same results for the same inputs, regardless of the timing of the calculations. This is crucial for ensuring the repeatability and reliability of the simulation.

10.2.2 Meeting Timing Deadlines:

Real-time simulations must meet strict timing deadlines. If the simulation fails to complete its calculations within the allotted time step, it will miss the deadline, leading to timing errors and potentially invalid simulation results.

10.2.3 Applications of Real-Time Simulation:

- **Human-in-the-Loop (HIL) Testing:** Allows human operators to interact with the simulation in real time, providing valuable feedback on the system's handling qualities and control system performance.
- **Hardware-in-the-Loop (HIL) Testing:** Integrates physical hardware components into the simulation loop, allowing for more realistic testing of the hardware and software interactions.
- **Rapid Prototyping:** Allows engineers to quickly prototype and test new control system designs in a real-time environment.
- **Pilot Training:** Provides a realistic and immersive training environment for pilots.

10.3 Real-Time Operating Systems (RTOS): The Foundation of Real-Time Performance

Real-Time Operating Systems (RTOS) are specialized operating systems designed to provide deterministic and predictable timing behavior. Unlike general-purpose operating systems, RTOSs prioritize real-time tasks and guarantee that these tasks will be executed within a specified time frame.

10.3.1 Key Features of RTOSs:

- **Preemptive Scheduling:** Allows high-priority tasks to interrupt lower-priority tasks, ensuring that critical tasks are executed promptly.
- **Priority Inversion Avoidance:** Techniques (e.g., priority inheritance, priority ceiling protocol) prevent priority inversion, a situation where a low-priority task blocks a high-priority task.
- **Real-Time Clock:** Provides a precise and accurate time base for synchronizing the simulation with the real world.
- **Memory Management:** Efficient memory management techniques minimize memory fragmentation and ensure predictable memory access times.

10.3.2 Examples of RTOSs:

- VxWorks
- QNX
- RTLinux
- FreeRTOS

10.3.3 Selecting an RTOS:

The choice of RTOS depends on the specific requirements of the application. Factors to consider include:

- **Determinism:** The required level of determinism.

- **Real-Time Performance:** The required real-time performance.
- **Cost:** The cost of the RTOS license.
- **Support:** The availability of technical support and documentation.

10.4 Input/output (I/O) Interfaces: Bridging the Gap

Real-time simulations often need to interact with the physical world, either to receive sensor data or to send control signals to actuators. This requires the use of specialized I/O interfaces.

10.4.1 Types of I/O Interfaces:

- **Analog-to-Digital Converters (ADCs):** Convert analog signals from sensors into digital data that can be processed by the simulation.
- **Digital-to-Analog Converters (DACs):** Convert digital control signals from the simulation into analog signals that can be used to control actuators.
- **Serial Communication Interfaces (e.g., UART, SPI, I2C):** Used to communicate with sensors and actuators that use serial communication protocols.
- **Ethernet:** Used for high-speed communication between the real-time computer and other devices.
- **MIL-STD-1553:** A military standard for communication between avionics systems.
- **ARINC 429:** A commercial standard for communication between avionics systems.
- **CAN Bus:** A communication protocol commonly used in automotive and aerospace applications.

10.4.2 Real-Time Considerations for I/O:

- **Latency:** The delay between the time a signal is generated and the time it is received. Minimizing latency is crucial for real-time performance.
- **Jitter:** The variation in latency. Minimizing jitter is important for ensuring the determinism of the simulation.
- **Throughput:** The rate at which data can be transferred. The I/O interface must have sufficient throughput to handle the data rates required by the simulation.

10.5 Hardware-in-the-Loop (HIL) Testing: Closing the Loop with Reality

Hardware-in-the-Loop (HIL) testing is a technique that integrates physical hardware components into the simulation loop. This allows for more realistic testing of hardware-software interactions and provides a powerful method to validate the design of complex aerospace systems.

10.5.1 Components of a HIL Simulation:

- **Real-Time Computer:** Runs the real-time simulation.
- **I/O Interfaces:** Connect the real-time computer to the hardware components.
- **Hardware Components:** The physical hardware components being tested (e.g., flight control computer, actuators, sensors).
- **Software Models:** Represent the remaining components of the system.
- **Signal Conditioning:** Hardware to scale signals, filter noise, and provide isolation between components.

10.5.2 Benefits of HIL Testing:

- **Increased Fidelity:** Provides a more realistic testing environment than purely software-based simulation.
- **Early Detection of Errors:** Helps detect hardware or software design errors early in the development process.
- **Reduced Risk:** Minimizes the risk of damaging expensive hardware during testing.
- **Cost Savings:** Reduces flight testing costs by allowing engineers to identify and fix problems before actual flight tests.

10.5.3 Types of HIL Testing:

- **Open-Loop HIL:** The hardware component is stimulated with pre-defined inputs, and its response is measured.
- **Closed-Loop HIL:** The hardware component is integrated into a feedback control loop, and its behavior is evaluated in response to real-time simulation outputs.
- **Software-in-the-Loop (SIL):** The hardware is replaced with a software model, allowing early testing of the software before hardware availability.
- **Processor-in-the-Loop (PIL):** The software runs on the actual processor that will be used in the hardware, allowing testing of the software's performance on the target hardware.

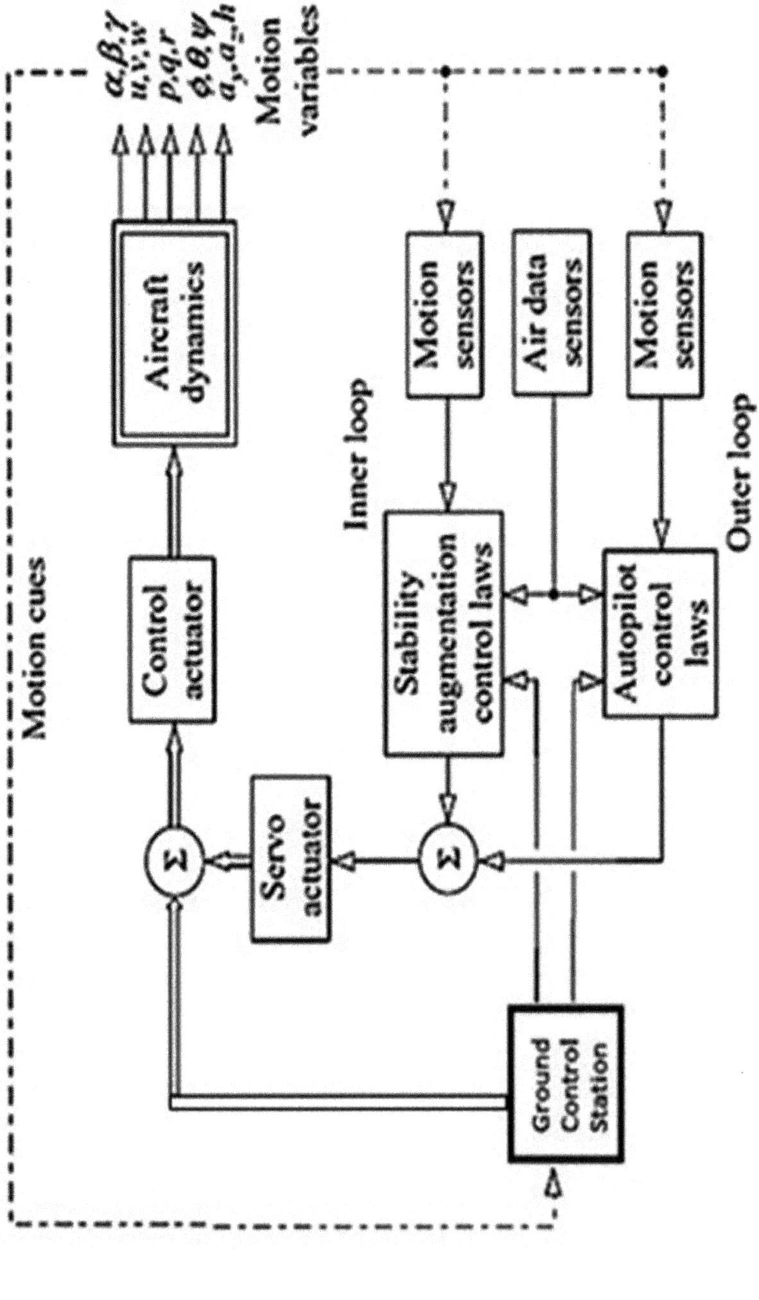
Motion cues
Control actuator
Aircraft dynamics
α,β,γ
u,v,w
p,q,r
φ,θ,ψ
a_y,a_z,h
Motion variables
Inner loop
Motion sensors
Air data sensors
Motion sensors
Outer loop
Stability augmentation control laws
Autopilot control laws
Servo actuator
Σ
Σ
Ground Control Station

10.6 Modeling Hardware Components for HIL Testing

Accurate modeling of hardware components is crucial for the success of HIL testing. The models should capture the key characteristics of the hardware, including dynamics, nonlinearities, and operational limitations.

10.6.1 Modeling Actuators:

- **Transfer Function Models:** Represent the actuator's dynamics using a transfer function that relates the input signal to the output position or force.
- **Nonlinear Models:** Capture nonlinearities such as saturation, hysteresis, and deadband.
- **Fault Models:** Simulate potential faults, including stuck valves, broken gears, or electrical failures.

10.6.2 Modeling Sensors:

- **Transfer Function Models:** Represent sensor dynamics using a transfer function relating the input quantity to the output signal.
- **Noise Models:** Simulate inherent sensor noise.
- **Bias Models:** Simulate systematic offsets in sensor readings.
- **Fault Models:** Simulate potential faults such as drift, saturation, or complete sensor failure.

10.6.3 Example: Modeling an Actuator with Saturation

Actuators have output limits, which can be modeled as a saturation block. For example, an actuator with a maximum force of 100 N and a minimum force of -100 N can be modeled in Python as:

Python;

```
def actuator_model(command):
max_force = 100
min_force = -100
force = command  # Ideal actuator output
# Apply saturation limits
if force > max_force:
force = max_force
elif force < min_force:
force = min_force
return force
```

This simple model ensures that the actuator output remains within its physical limits, which is critical for realistic HIL simulation.

10.7 Overcoming Latency and Jitter Challenges

Latency and jitter can significantly degrade the accuracy and stability of real-time simulations. The following techniques can be employed to mitigate these effects:

10.7.1 Reducing Computational Latency:

- **Optimize Code:** Streamline the simulation code to minimize execution time.
- **Use Efficient Algorithms:** Employ computationally efficient numerical integration methods and control algorithms.
- **Parallel Processing:** Distribute computational load across multiple processors or cores to speed up calculations.

10.7.2 Reducing I/O Latency:

- **High-Performance I/O Interfaces:** Utilize interfaces with low latency and minimal jitter.
- **Direct Memory Access (DMA):** Transfer data directly between I/O devices and memory without involving the CPU.
- **Time-Stamping:** Time-stamp sensor data to account for any latency in the I/O system and ensure correct sequencing.

10.7.3 Compensating for Time Delays:

- **Predictive Filters:** Use filters like Kalman filters to estimate future system states based on past measurements.
- **Smith Predictor:** Implement a Smith Predictor control scheme to compensate for known delays in feedback loops, improving stability and response.

These strategies collectively help maintain determinism, improve accuracy, and ensure stable operation in real-time and HIL simulations.

10.8 Validation of Real-Time Simulations

Validating real-time simulations is essential to ensure that they accurately represent the behavior of the real-world system. Validation involves comparing simulation results with experimental or flight test data and assessing system performance under various conditions.

10.8.1 Key Validation Metrics:

- **Timing Accuracy:** Ensure that the simulation maintains correct real-time timing and meets all deadlines.
- **Stability:** Verify that the simulation remains stable and does not exhibit unbounded oscillations or divergence.

- **Accuracy of Hardware Response:** Compare the hardware's response against the simulation using metrics such as frequency response, phase, and damping ratio.

10.8.2 Techniques for Validation:

- **Comparison with Experimental Data:** Compare simulation outputs with results from bench tests, flight tests, or operational data to assess fidelity.
- **Sensitivity Analysis:** Evaluate how variations in model parameters affect the simulation results, identifying critical parameters and model robustness.
- **Fault Injection Tests:** Introduce faults or abnormal conditions in the simulation to verify that the hardware and control system detect, isolate, and respond appropriately.

Proper validation ensures that the real-time simulation is both reliable and predictive, providing confidence for HIL testing, control system design, and pilot-in-the-loop applications.

10.9 Case Study: HIL Testing of a Flight Control Computer (FCC)

Hardware-in-the-Loop (HIL) testing provides a realistic environment to validate and verify the performance of a flight control computer (FCC) before actual flight. This case study illustrates a typical HIL testing setup for an aircraft FCC.

10.9.1 Setup

- **Real-Time Computer:** Runs a real-time simulation of the aircraft's dynamics, aerodynamics, and environmental conditions.
- **I/O Interfaces:** Connect the real-time computer to the FCC and the actuators, enabling bidirectional communication.
- **Flight Control Computer (FCC):** The physical FCC under test.

- **Actuators:** Simulated actuators that respond to control signals generated by the FCC.
- **Sensors:** Simulated sensors providing real-time feedback of the aircraft state to the FCC.

10.9.2 Testing Procedure

1. Connect the FCC to the real-time computer along with the simulated actuators and sensors.
2. Start the real-time simulation of the aircraft.
3. The FCC receives sensor data from the simulation and computes control commands.
4. Control commands are sent to the simulated actuators, which respond accordingly.
5. The aircraft's state is updated in the simulation based on the actuator outputs.
6. Updated sensor data reflecting the new aircraft state is sent back to the FCC.
7. Steps 3–6 repeat continuously in real-time, closing the loop between the FCC and the simulated environment.

10.9.3 Tests Performed

- **Step Response Testing:** Verify tracking performance, stability, and control authority.
- **Scenario Testing:** Run realistic flight scenarios in the HIL environment and compare results to offline simulations.
- **Extended Automated Testing:** Conduct thousands of hours of automated tests to identify unexpected behaviors or system crashes.
- **Fault Injection Testing:** Introduce actuator or sensor failures to ensure the FCC can detect faults and respond appropriately.

HIL testing in this manner allows engineers to evaluate FCC performance safely, identify design issues early, and reduce the risk and cost associated with actual flight testing.

10.10 Emerging Trends in Real-Time and HIL Simulation

- **Cloud-Based HIL:** Moving HIL simulations to the cloud enables scalable, remote testing of aerospace systems, allowing multiple teams to collaborate without being co-located.
- **AI-Enhanced HIL:** Artificial intelligence can automatically generate test cases, detect anomalies, and analyze simulation results, improving testing efficiency and coverage.
- **More Realistic Environments:** Advances in computational power allow higher-fidelity simulations of complex phenomena, including turbulence, engine failures, sensor noise, and other environmental effects.

10.11 Conclusion

Real-time simulation and Hardware-in-the-Loop (HIL) testing are essential methodologies for the development, validation, and verification of modern aerospace systems. By integrating physical hardware into a simulation loop, HIL provides a highly realistic and reliable testing environment that surpasses purely software-based simulations. As aerospace systems grow more complex, real-time simulation and HIL testing will continue to play a pivotal role in ensuring system safety, performance, and robustness.

10.12 Exercises

1. Define real-time simulation and explain its importance in aerospace systems.
2. Describe the key features of a Real-Time Operating System (RTOS) and their significance.
3. List and explain the different types of I/O interfaces used in real-time simulations.
4. What is Hardware-in-the-Loop (HIL) testing, and what benefits does it provide over traditional simulations?
5. Explain how actuators and sensors are modeled for HIL testing, including considerations for nonlinearities, noise, and faults.
6. Discuss the challenges of achieving real-time performance, such as latency and jitter, and describe methods to overcome them.
7. Design a HIL simulation setup for testing a UAV autopilot, including real-time computer, I/O interfaces, sensors, actuators, and testing procedures.

11.2 Sources of Uncertainty: Tracing the Roots of Imperfection

Understanding the sources of uncertainty is the first step in effectively managing it. These sources can be broadly categorized as:

11.2.1 Model Uncertainty:

Arises from the simplifications and approximations made when developing a mathematical model of the system. This can include:

- **Structural Uncertainty:** Uncertainty in the form or structure of the model itself (e.g., neglecting higher-order aerodynamic effects).
- **Parametric Uncertainty:** Uncertainty in the values of the parameters used in the model (e.g., aerodynamic coefficients, engine thrust parameters, mass properties).
- **Environmental Uncertainty:** Uncertainty in the environmental conditions (e.g., wind gusts, atmospheric density variations).

11.2.2 Data Uncertainty:

Arises from the errors and limitations in the data used to calibrate and validate the model. This can include:
• **Measurement Error:** Errors in the sensors used to collect data (e.g., noise, bias, drift).

- **Statistical Uncertainty:** Uncertainty due to limited sample size or incomplete data.
- **Calibration Errors:** Errors in the process of calibrating the model to match experimental data.

11.2.3 Numerical Uncertainty:

Arises from the limitations of the numerical methods used to solve the equations of motion. This can include:

- **Discretization Error:** Errors due to approximating continuous equations with discrete numerical methods.
- **Round-off Error:** Errors due to the finite precision of computers.
- **Convergence Error:** Errors due to the iterative nature of some numerical methods (e.g., optimization algorithms).

11.2.4 Human Factors Uncertainty:

Uncertainties related to human input or operation, such as variations in pilot skill or errors in entering commands.

11.3 Uncertainty Quantification (UQ): Measuring the Unknown

Uncertainty Quantification (UQ) is the process of characterizing and quantifying the uncertainty in the simulation results due to the various sources of uncertainty in the model, data, and numerical methods.

11.3.1 Probabilistic UQ:

Assumes that the uncertain parameters can be described by probability distributions. The goal is to propagate these probability distributions through the simulation to obtain probability distributions for the simulation outputs.

- **Monte Carlo Simulation (MCS):**

A widely used technique for probabilistic UQ. MCS involves running the simulation multiple times with different sets of random samples drawn from the probability distributions of the uncertain parameters. The results of the multiple simulations are then used to estimate the probability distributions of the simulation outputs.

• ***(Calculation Example: Monte Carlo Simulation)***

Suppose we want to estimate the uncertainty in the range of a projectile due to uncertainty in the initial velocity. Assume the initial velocity has a normal distribution with a mean of 100 m/s and a standard deviation of 5 m/s. We can use MCS to run the simulation multiple times with different values of the initial velocity drawn from this distribution. The range of the projectile is then calculated for each simulation run, and the results are used to estimate the mean, standard deviation, and other statistics of the range.

Python; Run

```
import numpy as np
import matplotlib.pyplot as plt
# Define the parameters of the initial velocity distribution
mean_velocity = 100  # m/s
std_velocity = 5    # m/s
# Number of Monte Carlo samples
num_samples = 1000
# Generate random samples from the normal distribution
velocities = np.random.normal(mean_velocity, std_velocity, num_samples)
# Define the simulation parameters (simplified projectile motion)
gravity = 9.81      # m/s^2
angle = np.radians(45)  # Launch angle in radians
# Function to calculate the range of the projectile
def calculate_range(velocity, angle, gravity):
    return (velocity**2 * np.sin(2*angle)) / gravity
# Calculate the range for each velocity sample
ranges = calculate_range(velocities, angle, gravity)
# Calculate statistics of the range
```

```
mean_range = np.mean(ranges)
std_range = np.std(ranges)
# Display the results
print(f"Mean Range: {mean_range:.2f} m")
print(f"Standard Deviation of Range: {std_range:.2f} m")
# Plot a histogram of the range values
plt.hist(ranges, bins=30, density=True)
plt.xlabel("Range (m)")
plt.ylabel("Probability Density")
plt.title("Distribution of Projectile Range (Monte Carlo Simulation)")
plt.show()
```

- **Polynomial Chaos Expansion (PCE):**

Approximates the simulation output as a polynomial function of the uncertain parameters. PCE can be more efficient than MCS for certain types of problems,but it requires a smooth and well behaved simulation output.

- **Latin Hypercube Sampling (LHS):**

An improved sampling technique over basic Monte Carlo that better populates the parameter space with fewer samples.

11.3.2 Non-Probabilistic UQ:

Does not assume that the uncertain parameters can be described by probability distributions. Instead, it uses set-based methods to represent the uncertainty, such as intervals or ellipsoids. The goal is to determine the range of possible simulation outputs given the set-based representation of the uncertain parameters.

- **Interval Analysis:** Represents the uncertain parameters as intervals. The simulation is then run with the endpoints of the intervals to determine the range of possible simulation outputs.

- **Affinity Arithmetic:** A more sophisticated technique that tracks the correlations between the uncertain parameters and the simulation outputs.

11.4 Sensitivity Analysis (SA): Identifying the Influential Factors

Sensitivity Analysis (SA) is the process of identifying and quantifying the influence of different input parameters on the simulation outputs. SA helps to determine which parameters have the greatest impact on the simulation results and which parameters can be fixed at nominal values without significantly affecting the accuracy of the simulation.

11.4.1 Local Sensitivity Analysis:

Examines the sensitivity of the simulation outputs to small changes in the input parameters around a nominal operating point.

• **One-at-a-Time (OAT) Method:** Varies each input parameter one at a time while holding the other parameters constant. The sensitivity is then calculated as the change in the output divided by the change in the input.

• ***(Calculation Example: OAT Sensitivity Analysis)*** suppose we want to determine the sensitivity of the range of a projectile to the initial velocity. We can vary the initial velocity by a small amount (e.g., 1 m/s) while holding the other parameters constant. The sensitivity is then calculated as the change in the range divided by the change in the initial velocity.

Sensitivity = (Range(101 m/s) - Range(100 m/s)) / (101 m/s - 100 m/s)

Sensitivity = (1040.82 m - 1029.59 m) / (1 m/s) = 11.23 m/(m/s)

(assuming Gravity = 9.81 m/s^2, Angle = 45 degrees)

• **Gradient-Based Methods:** Calculate the gradient of the simulation output with respect to the input parameters. The gradient provides a measure of the sensitivity of the output to each parameter.

11.4.2 Global Sensitivity Analysis:

Examines the sensitivity of the simulation outputs to variations in the input parameters over their entire range of possible values.

• **Variance-Based Methods:** Decompose the variance of the simulation output into contributions from each input parameter and their interactions. This provides a measure of the relative importance of each parameter.

• **Sobol Indices:** A widely used type of variance-based method. Sobol indices quantify the fraction of the total variance of the simulation output that is due to each input parameter and their interactions.

• **(Calculation Example: Sobol Indices (Conceptual))** Imagine we have two uncertain parameters, A and B, affecting a simulation output Y. After performing a global sensitivity analysis, we obtain the following Sobol indices:

• First-Order Sobol Index for A (S_A): 0.6

• First-Order Sobol Index for B (S_B): 0.2

• Total-Order Sobol Index for A (ST_A): 0.7

• Total-Order Sobol Index for B (ST_B): 0.3

This indicates:

* Parameter A is more influential than parameter B (S_A > S_B).

* Parameter A has some interactions with other parameters (ST_A > S_A).

* Parameter B has minimal interactions with other parameters (ST_B ≈ S_B).

• **Morris Method (Elementary Effects Test):** A computationally efficient method that samples the input parameter space and calculates the elementary effect of each parameter on the simulation output.

11.5 Combining UQ and SA: A Powerful Approach

UQ and SA are complementary techniques that can be used together to provide a comprehensive assessment of the reliability and robustness of aerospace vehicle dynamics simulations

• **Steps:**

1. Identify the Uncertain Parameters: Identify all of the sources of uncertainty in the simulation model, data, and numerical methods.
2. Quantify the Uncertainties: Quantify the uncertainties by assigning probability distributions or set-based representations to the uncertain parameters.
3. Perform Uncertainty Quantification: Use UQ techniques to propagate the uncertainties through the simulation and obtain probability distributions or ranges for the simulation outputs.
4. Perform Sensitivity Analysis: Use SA techniques to identify the input parameters that have the greatest impact on the simulation outputs.

5. Reduce Uncertainty: Focus on reducing the uncertainty in the most influential parameters. This may involve collecting more data, improving the model, or using more accurate numerical methods.
6. Re-validate and Refine: Re-validate the simulation with the reduced uncertainties and refine the model or simulation parameters as needed.

11.6 Applications in Aerospace Vehicle Dynamics

UQ and SA have numerous applications in aerospace vehicle dynamics:

11.6.1 Robust Control System Design:

Design control systems that are robust to uncertainties in the aircraft dynamics, sensor noise, and actuator limitations.

11.6.2 Flight Envelope Protection:

Develop flight envelope protection systems that prevent the aircraft from exceeding its operational limits, even in the presence of uncertainties.

11.6.3 Mission Planning:

Plan missions that are robust to uncertainties in the weather conditions, fuel consumption, and other factors.

11.6.4 Risk Assessment:

Assess the risks associated with operating aerospace vehicles in uncertain environments.

11.6.5 Verification and Validation (V&V):

Increase confidence that the simulation is fit for purpose.

11.7 Example: Robust Controller Design Using UQ and SA

Consider designing a pitch rate controller for an aircraft. There is uncertainty in the aerodynamic derivatives.

1. **Identify Uncertain Parameters:** The parameters with uncertainty are the stability derivatives C_{mα} and C_{mδe} (pitching moment due to angle of attack and elevator deflection, respectively).
2. **Quantify Uncertainties:** Represent the uncertainty in C_{mα} and C_{mδe} using normal distributions with means equal to their nominal values and standard deviations based on experimental data.
   C_{mα} ~ N(-0.8, 0.1)
   C_{mδe} ~ N(-1.2, 0.15)
3. **Perform UQ and SA:**
   • Use Monte Carlo simulation to generate multiple sets of values for C_{mα} and C_{mδe} based on the distributions.
   • For each set of parameters, design a PID controller to achieve desired performance (e.g., settling time, overshoot).
   • Calculate the gain margins and phase margins of the closed-loop system for each set of parameters.
   • Plot the regions of stability in gain and phase margins.
4. **Design a Robust Controller:** Select the controller gain that keeps the system stable over the range of likely parameter values.

11.8 Emerging Trends and Future Directions

11.8.1 Machine Learning for UQ and SA:

Machine learning techniques can be used to develop more efficient and accurate UQ and SA methods.

11.8.2 Multi-Fidelity UQ:

Combine high-fidelity and low-fidelity simulations to reduce the computational cost of UQ.

11.8.3 Uncertainty-Aware Design:

Design aerospace vehicles and control systems that explicitly account for uncertainty in the design process.

- **Model Calibration** to reduce uncertainties.

11.9 Conclusion

Uncertainty Quantification and Sensitivity Analysis are essential tools for building confidence in the results of aerospace vehicle dynamics simulations. By understanding the sources of uncertainty and using appropriate techniques to quantify and analyze their impact, engineers can design more robust and reliable aerospace systems. As aerospace systems become increasingly complex and operate in more challenging environments, the importance of UQ and SA will only continue to grow.

11.10 Exercises

1. Explain the difference between aleatory and epistemic uncertainty.
2. Describe the key steps in performing a Monte Carlo simulation.

3. What are Sobol indices, and how can they be used to assess the sensitivity of simulation outputs to different input parameters?
4. Design a robust control system for an aircraft that is robust to uncertainties in the aerodynamic derivatives.
5. Analyze the impact of wind gusts on the trajectory of a drone using UQ techniques.
6. What strategies would you use to account for large but rare environmental events?
7. Research Bayesian methods for model calibration.
8. How can you effectively validate a UQ process?

ANSWERS

Here are suggested answers to the exercises posed at the end of Chapters 2 through 11. Keep in mind that some of these are more open-ended and require research or creative problem-solving, so the answers provided are examples or starting points.

Chapter 2: Coordinate Systems and Transformations

1. **Explain the purpose of a coordinate system in aerospace vehicle dynamics.**
   • A coordinate system provides a reference frame to define the position and orientation of objects in space with numerical values (coordinates). This allows us to describe and analyze motion quantitatively, making concepts like velocity and acceleration meaningful. Without a defined frame, these concepts are ambiguous.

2. **Describe the differences between the inertial frame and the Earth-Centered Inertial (ECI) frame.**
   • The Inertial Frame is a theoretical, non-accelerating, non-rotating reference frame "fixed" in space and used as a universal reference for motion. A true inertial frame is an idealization. The ECI frame is centered at the Earth's center of mass and its axes point to fixed stars. It rotates slightly as the Earth orbits the sun and is, therefore, only an approximation of an inertial frame. The ECI frame is useful for Earth-based aerospace calculations.

3. **What are Euler angles? Explain the Roll-Pitch-Yaw sequence and why the order matters.**
   • Euler angles are a set of three angles that describe the orientation of one coordinate system relative to another through a sequence of rotations. The Roll-Pitch-Yaw

sequence (ϕ, θ, ψ) is a rotation of ϕ about the x-axis (roll), followed by θ about the y-axis (pitch), and followed by ψ about the z-axis (yaw). The order matters because rotation matrices are not commutative, meaning the final orientation depends on the sequence of rotations. Changing the order will result in a different final orientation.

4. **Why are quaternions sometimes preferred over Euler angles for representing rotations?**
   • Quaternions are preferred because they avoid gimbal lock (a loss of one degree of freedom when two axes align), provide a more compact representation (four numbers instead of nine), and are better suited for interpolating between rotations smoothly.

5. **Write a Python function that takes Euler angles (in degrees) as input and returns the corresponding rotation matrix (using the Roll-Pitch-Yaw sequence).**
   Python:

```
import numpy as np

def euler_to_rotation_matrix(phi_deg, theta_deg, psi_deg):
    phi = np.radians(phi_deg)
    theta = np.radians(theta_deg)
    psi = np.radians(psi_deg)

    Rx = np.array([[1, 0, 0], [0, np.cos(phi), -np.sin(phi)], [0, np.sin(phi), np.cos(phi)]])
    Ry = np.array([[np.cos(theta), 0, np.sin(theta)], [0, 1, 0], [-np.sin(theta), 0, np.cos(theta)]])
    Rz = np.array([[np.cos(psi), -np.sin(psi), 0], [np.sin(psi), np.cos(psi), 0], [0, 0, 1]])

    R = Rz @ Ry @ Rx
    return R
```

```
# Example usage:
phi = 10
theta = 5
psi = 2
rotation_matrix = euler_to_rotation_matrix(phi, theta, psi)
print(rotation_matrix)
```

6. An aircraft is flying with a heading of 45 degrees (yaw), a pitch angle of 10 degrees, and a roll angle of 2 degrees. A sensor on the aircraft measures an acceleration of 1,0,01, 0, 01,0,0 m/s^2 in the body frame. What is the acceleration vector in the inertial frame? Provide your code (Python preferred) and the result.

7. **Research and describe one real-world scenario where accurately transforming between different coordinate systems is essential for a successful aerospace mission.**
   • Satellite Rendezvous and Docking: Accurately transforming between the chaser satellite's body frame, the target satellite's body frame, and an inertial frame is critical for successful rendezvous and docking. Precise relative position and attitude information are needed for navigation and control to avoid collisions and ensure a successful docking maneuver. Errors in coordinate transformations can lead to missed docking attempts or even catastrophic collisions.

Chapter 3: Equations of Motion

1. **Explain the difference between a force and a moment. Give examples of each that are relevant to aerospace vehicles.**
   • A force is a vector quantity that tends to cause linear

acceleration of a body. Examples in aerospace: lift, drag, thrust, gravity.

• A moment (or torque) is a vector quantity that tends to cause angular acceleration of a body. Examples: rolling moment, pitching moment, yawing moment, control surface moments.

2. **Derive the linear equations of motion for a simplified case where the vehicle is only allowed to move in one dimension (e.g., vertical motion).**

   • Assume vertical motion is along the z-axis.

   • Forces: Gravity (mg, downward), Thrust (T, upward), Drag (D, downward, $D = 0.5 * \rho * V^2 * A * C_D$).

   • Net Force: $F_{net} = T - mg - D$.

   • Newton's Second Law: $F_{net} = ma$.

   • Linear Equation of Motion: $ma = T - mg - D \Rightarrow a = (T - mg - D) / m$.

   • Therefore: $z'' = (T - mg - (0.5 * \rho * z'^2 * A * C_D)) / m$ (where z'' is vertical acceleration and z' is vertical velocity).

3. **What is the inertia tensor? Explain its significance in the angular equations of motion.**

   • The inertia tensor is a 3×3 matrix that describes the distribution of mass within a body and its resistance to angular acceleration. It relates the applied moments to the resulting angular accelerations. Its components are moments of inertia (resistance to rotation around a single axis) and products of inertia (represent coupling between rotations about different axes).

• The inertia tensor appears in the angular equations of motion: $\mathbf{M} = \mathbf{I}\,\boldsymbol{\alpha} + \boldsymbol{\omega} \times (\mathbf{I}\boldsymbol{\omega})$.

• It accounts for how the vehicle's mass distribution affects its rotational dynamics.

4. **Research and describe a real-world scenario where gyroscopic moments have a significant impact on the behavior of an aerospace vehicle.**

   • Space Station Attitude Control: Large rotating components, like solar arrays, on the International Space Station (ISS) can generate significant gyroscopic torques when the station changes its attitude. These torques must be carefully managed by the attitude control system to maintain stability and prevent unwanted oscillations.

5. **An aircraft has an inertia tensor with I_{xx} = 5000 kg·m², I_{yy} = 8000 kg·m², and I_{zz} = 10000 kg·m². It is rolling at a rate of p = 0.2 rad/s and pitching at a rate of q = 0.1 rad/s. Assuming that the products of inertia are negligible, calculate the yawing moment generated by the gyroscopic effect.**

   • $N = (I_{yy} - I_{xx}) * p * q$

   • $N = (8000\ \text{kg·m}^2 - 5000\ \text{kg·m}^2) * 0.2\ \text{rad/s} * 0.1\ \text{rad/s}$

   • $N = 3000 * 0.02 = 60\ \text{N·m}$

Chapter 4: Aerodynamic Modeling

1. **Explain the difference between lift and drag.**

   • Lift is the aerodynamic force acting perpendicular to the relative wind, primarily supporting the vehicle's weight and enabling flight.

• Drag is the aerodynamic force acting parallel to the relative wind, opposing the vehicle's motion through the air.

2. **What are aerodynamic coefficients, and why are they useful?**

   • Aerodynamic coefficients are dimensionless numbers that relate aerodynamic forces and moments to the vehicle's geometry, airspeed, and air density (e.g., C_l, C_d, C_m).

   • They are useful because they:

   • Characterize aerodynamic performance independent of specific size or flight conditions.

   • Allow comparison of different designs.

   • Simplify aerodynamic calculations by encapsulating complex relationships.

3. **Describe the key factors that affect the aerodynamic performance of a wing.**

   • Angle of attack: Influences lift and drag.

   • Airspeed: Aerodynamic forces are proportional to airspeed squared.

   • Air density: Influences aerodynamic force magnitude.

   • Wing area: Directly affects lift and drag forces.

   • Aspect ratio: Affects induced drag.

   • Wing planform: Shape influences lift distribution and stall characteristics.

   • Airfoil shape: Determines lift and drag characteristics.

   • Control surface deflections.

4. **What is induced drag, and how can it be reduced?**

   • Induced drag is drag caused by the creation of wingtip vortices due to finite wingspan. These vortices create downwash, which reduces the effective angle of attack.

   • It can be reduced by:

   • Increasing aspect ratio (longer, slender wings).

   • Using winglets.

   • Employing an elliptical wing planform.

5. **Research and compare the advantages and disadvantages of different aerodynamic modeling techniques (look-up tables, empirical models, VLM, CFD).**

Technique	Advantages	Disadvantages
Look-up Tables	Computationally efficient, simple to implement.	Requires extensive experimental or CFD data; may not capture complex phenomena; no extrapolation possible.
Empirical Models	More flexible than look-up tables; can extrapolate.	Less accurate than CFD for complex flows; relies on empirical data.
VLM	Efficient for lift and induced drag; handles complex wing geometries.	Not suitable for viscous flows; limited accuracy for complex flows.
CFD	Most accurate; captures complex	Computationally expensive; requires CFD expertise.

	phenomena (turbulence, stall).	

6. **An aircraft is flying at an airspeed of 150 m/s at an altitude of 5000 m (ρ = 0.736 kg/m³). The wing area is 25 m², and the lift coefficient is C_l = 0.6. Calculate the lift force acting on the aircraft.**

   • $L = 0.5 * \rho * V^2 * S * C_l$

   • $L = 0.5 * 0.736\ kg/m^3 * (150\ m/s)^2 * 25\ m^2 * 0.6$

   • L = 62,100 N

7. **Using an online airfoil database (e.g., Airfoil Tools), research the characteristics of different airfoil shapes. Compare and contrast the lift and drag performance of a symmetric airfoil with that of a cambered airfoil.**

   • **Symmetric Airfoil:** Generates zero lift at zero angle of attack; lower drag at low angles of attack; stall angle is the same for positive and negative AoA. Suitable for aircraft that spend significant time inverted.

   • **Cambered Airfoil:** Generates lift at zero angle of attack; higher maximum lift coefficient; higher drag at low angles of attack; different stall angles for positive and negative AoA. Suitable for efficient cruising

Chapter 5: Propulsion System Modeling

1. **Explain the basic principle of thrust generation.**

   • Thrust is generated by accelerating a mass of working fluid (typically air or exhaust gases) in one direction. The reaction force to this acceleration propels the vehicle in the opposite direction (Newton's Third Law).

2. **Describe the differences between turbojet, turbofan, and turboprop engines.**

    • **Turbojet:** All air passes through the core engine; high exhaust velocity; efficient at high speeds.

    • **Turbofan:** A large fan bypasses some air around the core; higher thrust and fuel efficiency at lower speeds compared to turbojets.

    • **Turboprop:** Turbine drives a propeller for primary thrust; most efficient at low speeds and altitudes.

3. **What is specific impulse, and why is it important for rocket engines?**

    • Specific impulse (I_{sp}) is a measure of a rocket engine's efficiency, representing the thrust produced per unit weight flow rate of propellant (measured in seconds).

    • It is important because a higher I_{sp} means more thrust generated per unit of propellant, leading to longer burn times and higher delta-V capabilities (the overall change in velocity achievable).

4. **Research and compare the advantages and disadvantages of different propulsion system modeling techniques (look-up tables, empirical models, physics-based models).**

Technique	Advantages	Disadvantages
Look-up Tables	Simple, computationally efficient, suitable for real-time simulation.	Requires pre-computed data; limited extrapolation capability; does not capture complex dynamics.
Empirical Models	More flexible than look-up tables; allows some extrapolation; computationally efficient.	Accuracy limited by the quality of empirical data; may not capture complex physics.
Physics-Based	Most accurate; captures complex physics; can be used for design and analysis.	Computationally expensive; requires detailed knowledge of engine components; more complex to implement.

5. **An aircraft engine produces a thrust of 30,000 N with a specific fuel consumption of 0.000025 kg/N/s. Calculate the fuel consumption rate in kg/hr.**

• Fuel consumption rate (kg/s) = SFC * Thrust = 0.000025 kg/N/s * 30,000 N = 0.75 kg/s

• Fuel consumption rate (kg/hr) = 0.75 kg/s * 3600 s/hr = 2700 kg/hr

6. **A rocket engine has a chamber pressure of 5 MPa and a nozzle expansion ratio of 20. Research the performance characteristics of different rocket propellants and estimate the exhaust velocity and specific impulse of the engine using a suitable propellant combination.**

   • A suitable propellant combination (good performance, reasonably safe) is Liquid Oxygen (LOX) / Kerosene (RP-1).

   • Typical chamber temperature for LOX/RP-1: ~3500 K

   • Effective exhaust velocity (Ve) can be estimated using gas dynamics calculations and propellant characteristics, but a rough estimate is Ve = 2500–3000 m/s.

   • Specific impulse (I_{sp}) = Ve / g_0 ≈ 2500 m/s / 9.81 m/s^2 ≈ 255 s

7. **Design a basic control system for a turbofan engine, including sensors for measuring engine speed, turbine inlet temperature, and fuel flow, and actuators for controlling the fuel flow rate and nozzle area.**

   • A possible design could include two primary control loops:

   • **Core Speed Controller:** Use fuel flow rate actuator to set the N1 turbine speed.

   • **Turbine Inlet Temperature Controller:** Modulate the air-bleed valves to adjust the turbine temperature.
   • **Sensors:** N1 speed, turbine inlet temperature, and fuel flow rate to provide feedback.

Chapter 6: Control Systems

1. **Explain the difference between open-loop and closed-loop control.**

   • **Open-loop control:** The control action is independent of the output of the system. Simple but susceptible to disturbances.

   • **Closed-loop control:** The control action is based on the output of the system (feedback). More complex but more accurate and robust.

2. **What is stability, and why is it important for control systems?**

   • Stability refers to a system's ability to return to its desired state after a disturbance. A stable control system avoids unbounded oscillations or divergence.

   • It is important because unstable systems are uncontrollable and can lead to catastrophic failures.

3. **Describe the three control actions in a PID controller.**

   • **Proportional (P):** Control signal proportional to the error. Reduces error but might leave a steady-state error.
   • **Integral (I):** Control signal proportional to the integral of the error. Eliminates steady-state error but can cause oscillations.
   • **Derivative (D):** Control signal proportional to the derivative of the error. Improves response time and damping but can amplify noise.

4. **What is state-space control, and how does it differ from classical control techniques?**

• State-space control represents the system using state variables and matrix equations, providing a comprehensive view of the system's dynamics.

• It differs from classical control (e.g., PID) by:

• Handling multi-variable systems easily.

• Providing more direct control over the system's internal behavior (through state feedback).

• Enabling optimal control design (e.g., LQR).

5. **Research and compare the advantages and disadvantages of different control techniques (PID control, state feedback control, LQR control, adaptive control, robust control).**

Technique	**Advantages**	**Disadvantages**
PID Control	Simple, easy to implement, widely applicable.	Requires tuning, less effective for complex or nonlinear systems, struggles with MIMO.
State Feedback	Optimal pole placement, direct control of system behavior.	Requires full state knowledge, sensitive to model inaccuracies.
LQR Control	Optimal trade-off between performance and control effort.	Requires accurate model, computationally intensive, sensitive to uncertainties.

Adaptive Control	Adjusts to changes in the plant dynamics, good for uncertain environments.	Can be complex to implement, can be unstable if not properly designed.
Robust Control	Insensitive to uncertainties, guarantees stability over a range of conditions.	Can be conservative, may sacrifice performance in nominal conditions.

6. **Design a PID controller for a simple DC motor speed control system.**

    - Requires knowledge of the motor model.
    - Design the loops based on step response testing and PID tuning.
    - Consider a cascaded loop structure to meet speed and settling targets.

7. **Design a state feedback controller for a simple inverted pendulum system.**

    • This is a standard controls problem.

8. **Explore the limitations of using a PID controller on an inherently unstable system and how more advanced control methods might provide better outcomes.**

    • While a well-tuned PID controller can often stabilize an unstable system, its performance may be limited, especially under disturbances or parameter variations. PID controllers are not designed to explicitly handle the inherent instability, which can lead to potential oscillations, slow response times, or even instability in certain conditions.

- **More advanced control methods:**

- **State Feedback Control:** Using state feedback, one can directly place the poles of the closed-loop system at desired locations, ensuring stability and shaping the response. This provides more precise control over the system's dynamics compared to PID.

- **Linear Quadratic Regulator (LQR):** LQR designs an optimal controller by minimizing a cost function that balances performance and control effort. This method can handle multi-variable systems and provide a robust control solution.

- **Model Predictive Control (MPC):** MPC uses a model of the system to predict future behavior and optimize control inputs over a finite time horizon. This method can handle constraints and is particularly effective for complex systems with nonlinearities or time delays.

Chapter 7: Simulation Techniques and Validation

1. **Explain the importance of numerical integration in aerospace vehicle simulation.**

    - Numerical integration is essential for approximating the solutions to the differential equations that govern the motion of aerospace vehicles. These equations are often too complex to solve analytically, requiring numerical methods to step through time and calculate the system's state.

2. **Compare and contrast different numerical integration methods (Euler, RK4, variable step-size).**

Method	Accuracy	Stability	Computational Cost	Advantages	Disadvantages

Euler	Low	Poor	Low	Simple to implement.	Inaccurate, unstable, especially for stiff systems.
RK4	Medium	Good	Medium	Good balance of accuracy and computational cost.	Less accurate than variable step-size methods for systems with stiff dynamics.
Variable Step-Size	High	Good	Variable	Adaptable, efficient for systems with varying dynamics.	More complex to implement than fixed-step methods.

3. **What is simulation validation, and why is it important?**

   • Simulation validation is the process of determining the degree to which a simulation accurately represents the real world from the perspective of the intended uses of the simulation.

   • It is important because it:

   • Builds confidence in simulation results.

   • Identifies model inaccuracies and limitations.

   • Ensures that the simulation is fit for its intended purpose.

   • Mitigates the risks of design decisions based on inaccurate results.

4. **Describe the steps involved in building a reliable aerospace vehicle simulation.**

   • Define objectives.

   • Develop models.

   • Choose a simulation environment.

   • Implement models.

   • Integrate models.

   • Configure simulation.

   • Run simulation.

   • Analyze results.

   • Validate simulation.

   • Iterate and refine.

5. **What is HIL simulation, and what are its benefits?**

   • HIL simulation integrates physical hardware components into the simulation loop.

   • Benefits:

   • Increased fidelity.

   • Early error detection.

   • Reduced risk.

   • Cost savings.

6. **Write a Python function that implements the RK4 method for a simple differential equation.**

```
def rk4_step(f, x, t, h):
  k1 = h * f(x, t)
```

```
  k2 = h * f(x + k1/2, t + h/2)
  k3 = h * f(x + k2/2, t + h/2)
  k4 = h * f(x + k3, t + h)
  x_next = x + (k1 + 2*k2 + 2*k3 + k4) / 6
  return x_next
# Example usage:
def my_function(x, t):
  return -2*x  # Example: x-dot = -2x
x_initial = 1
t_initial = 0
step_size = 0.1
x_next = rk4_step(my_function, x_initial, t_initial, step_size)
print(x_next)
```

7. **Design a simulation to analyze the effect of wind gusts on the stability of an aircraft.**

   - Develop Aircraft Dynamics Model.
   - Create Flight Control System.
   - Develop Wind Gust Model: Dryden or Von Karman model.
   - Implement Simulation Environment.
   - Test Cases: Implement step responses, perform automated sweeps.
   - Perform a sensitivity analysis on wind intensity.

Chapter 8: Case Studies in Aerospace Vehicle Dynamics Simulation

These are more involved and don't have simple definitive answers. The key is to think critically about the assumptions and simplifications, and to identify potential improvements to the models.

1. **For each of the case studies presented in this chapter, identify the key assumptions and simplifications that were made in the models. How might these assumptions affect the accuracy of the simulation results?**
   • **Quadcopter:** Rigid body, simplified aerodynamics, ideal motors.
   Effects: Reduced accuracy in turbulent conditions; neglecting motor dynamics.

   • **Rocket:** Rigid body, spherical Earth, simple atmosphere.

   Effects: Neglecting wind shear and complex gravity gradients.

   • **Shuttle:** Aerodynamic look-up tables, constant gravity, simplified controls.

   Effects: Reduced the need to test non-standard conditions; more accurate simulation under standard test conditions.

2. **Choose one of the case studies and develop a more detailed model for one of the components (e.g., the engine model, the aerodynamic model, or the control system model).**

   • **Rocket Engine Model:** Implement the effects of throttling and thrust vectoring, which are non-linear.

3. **For each of the case studies, suggest additional validation data sources that could be used to improve the confidence in the simulation results.**

   • **Quadcopter:** Real-world measurements.

   • **Rocket:** Telemetry from previous launches.

   • **Shuttle:** Flight test data from space missions.

4. **Research a real-world aerospace accident or incident and analyze how simulation could have been used to prevent the event.**

   • **Air France Flight 447:** A more sophisticated stall model coupled with human-in-the-loop simulation and pilot training could have prepared pilots for these complex situations.

5. **Develop a simulation of a simple aircraft performing a coordinated turn. Analyze the effects of bank angle and airspeed on the turn radius and load factor.**

   • See equations in text to calculate turn radius and load factor.

Chapter 9: Emerging Trends and Future Directions

These are speculative and require research. There isn't a "right" answer, but a well-reasoned and researched one is valuable.

1. **Research and write a report on the current state of quantum computing and its potential applications in aerospace engineering.**

   • Requires research, but should include information on qubit technology, quantum algorithms, and potential breakthroughs in areas like materials science, optimization, and simulation.

2. **Develop a machine-learning-based surrogate model for a simple aerodynamic system.**

    • Involves using a dataset (synthetic or real) to train a machine learning model (e.g., neural network) to predict aerodynamic coefficients.

3. **Design a digital twin for a UAV delivery system.**

    • Requires defining the components of the digital twin: UAV model, sensor data streams, weather data integration, route optimization algorithms, and predictive maintenance capabilities.

4. **Explore the use of virtual reality for pilot training or maintenance and repair procedures.**

    • Involves researching VR systems for training and how to perform maintenance.

5. **Discuss the ethical considerations of using machine learning and AI in aerospace vehicle dynamics simulation.**

    • Concerns about data bias in some scenarios.

    • Concerns that machine learning models may act as "black boxes."

6. **What strategies would you use to account for large, but rare environmental events?**

    • Use high-performance computers.

    • Design to be robust to uncertainties.

    • Apply model calibration to reduce uncertainties.

7. **Research Bayesian methods for model calibration.**

    • Start by collecting data.

- Select a likely model.
- Perform calculations to create a distribution.

8. **How can you effectively validate a UQ process?**

    - Start with a large dataset.
    - Compare with analytical or experimental results.
    - Perform a Monte Carlo study.

Chapter 10: Real-Time Simulation and Hardware-in-the-Loop Testing

1. **What is real-time simulation, and why is it important for aerospace systems?**

    - Real-time simulation runs at the same rate as real-world time. It's crucial for human-in-the-loop testing, HIL testing, and testing real-time control systems, providing a realistic environment for evaluating performance and safety.

2. **Explain the key features of a Real-Time Operating System (RTOS).**

    - Preemptive scheduling, priority inversion avoidance, real-time clock, and efficient memory management.

3. **Describe the different types of I/O interfaces used in real-time simulations.**

    - ADCs, DACs, serial communication interfaces (UART, SPI, I2C), Ethernet, MIL-STD-1553, ARINC 429, CAN Bus.

4. **What is Hardware-in-the-Loop (HIL) testing, and what are its benefits?**

• HIL testing integrates physical hardware components into the simulation loop.

• Benefits: increased fidelity, early error detection, reduced risk, and cost savings.

5. **Explain how to model actuators and sensors for HIL testing.**

   • Actuators: transfer function models, non-linear models (saturation, hysteresis), and fault models are used to accurately capture actuator behavior.

   • Sensors: transfer function models, noise models, bias models, and fault models are used to capture sensor effects.

6. **What are the challenges of achieving real-time performance, and how can these challenges be overcome?**

   • The challenges include reducing latency, jitter, and computational burden. These may be addressed by optimizing the existing design and selecting alternative components.

7. **Design a HIL simulation setup for testing a UAV autopilot.**

   • The HIL setup would consist of a real-time computer, I/O interfaces, UAV sensors and actuators, and software models to create a simulated environment.

Chapter 11: Uncertainty Quantification and Sensitivity Analysis

1. **Explain the difference between aleatory and epistemic uncertainty.**

• Aleatory Uncertainty (irreducible): Inherent randomness and variability in the system or environment (e.g., manufacturing tolerances, wind gusts). Cannot be reduced by gathering more information.

• Epistemic Uncertainty (reducible): Uncertainty due to a lack of knowledge about the system or model (e.g., inaccurate model parameters, simplified assumptions). Can be reduced by collecting more data or improving the model.

2. **Describe the key steps in performing a Monte Carlo simulation.**

• Define probability distributions for uncertain parameters.

• Generate random samples from these distributions.

• Run the simulation multiple times, each time using a different set of samples.

• Analyze the results to estimate the probability distributions of the outputs.

3. **What are Sobol indices, and how can they be used to assess the sensitivity of simulation outputs to different input parameters?**

    • Sobol indices quantify the fraction of the total variance of the simulation output that is due to each input parameter (first-order indices) and their interactions (total-order indices). They are used to determine the relative importance of each parameter in contributing to the overall uncertainty in the output.

4. **Design a robust control system for an aircraft that is robust to uncertainties in the aerodynamic derivatives.**

• One technique is to design several sets of controllers to be robust to the variations in the aerodynamic derivatives.

• Then make use of techniques that reduce variation or uncertainty in the control design.

5. **Analyze the impact of wind gusts on the trajectory of a drone using UQ techniques.**

   • Model wind gusts as random processes.

   • Use Monte Carlo simulation or polynomial chaos expansion to propagate the wind gust uncertainties through the drone's dynamics model.

   • Analyze the resulting probability distribution of the drone's position and orientation.

www.ingramcontent.com/pod-product-compliance
Lightning Source LLC
LaVergne TN
LVHW050933210926
848482LV00013B/599
* 9 7 9 8 2 9 5 5 7 5 2 8 0 *